# Genetics f

**The Remedica Genetics for... Series**
**Genetics for Cardiologists**
**Genetics for Dermatologists**
**Genetics for Endocrinologists**
**Genetics for Hematologists**
**Genetics for Oncologists**
**Genetics for Ophthalmologists**
**Genetics for Orthopedic Surgeons**
**Genetics for Pulmonologists**
**Genetics for Rheumatologists**

Published by the Remedica Group

Remedica Publishing Ltd, 32–38 Osnaburgh Street, London, NW1 3ND, UK

Remedica Inc, Tri-State International Center, Building 25, Suite 150, Lincolnshire, IL 60069, USA

E-mail: books@remedica.com

www.remedica.com

Publisher: Andrew Ward

In-house editor: Cath Harris

ISBN 1 901346 18 8

ISSN 1472 4618

Printed in Canada

British Library Cataloguing-in Publication Data

A catalogue record for this book is available from the British Library

# Genetics for Endocrinologists
# The molecular genetic basis
# of endocrine disorders

**Benjamin Glaser, MD**
Endocrinology and Metabolism Service
Internal Medicine Department
Hebrew University-Hadassah Medical School
Jerusalem
Israel

**Harry J Hirsch, MD**
Institute of Hormone Research
Shaare Zedek Medical Center
Jerusalem
Israel

Series Editor
**Eli Hatchwell**
Investigator
Cold Spring Harbor Laboratory

REMEDICA
p u b l i s h i n g

LONDON • CHICAGO

# Dedication

To our wives, Sara and Sharon, without whose constant love, support, and encouragement this book would not have been possible.

*Benjamin Glaser and Harry Hirsch*

# Introduction to Genetics for... series

Medicine is changing. The revolution in molecular genetics has fundamentally altered our notions of disease etiology and classification, and promises novel therapeutic interventions. Standard diagnostic approaches to disease focused entirely on clinical features and relatively crude clinical diagnostic tests. Little account was traditionally taken of possible familial influences in disease.

The rapidity of the genetics revolution has left many physicians behind, particularly those whose medical education largely preceded its birth. Even for those who might have been aware of molecular genetics and its possible impact, the field was often viewed as highly specialist and not necessarily relevant to everyday clinical practice. Furthermore, while genetic disorders were viewed as representing a small minority of the total clinical load, it is now becoming clear that the opposite is true: few clinical conditions are totally without some genetic influence.

The physician will soon need to be as familiar with genetic testing as he/she is with routine hematology and biochemistry analysis. While rapid and routine testing in molecular genetics is still an evolving field, in many situations such tests are already routine and represent essential adjuncts to clinical diagnosis (a good example is cystic fibrosis).

This series of monographs is intended to bring specialists up to date in molecular genetics, both generally and also in very specific ways that are relevant to the given specialty. The aims are generally two-fold:

(i)     to set the relevant specialty in the context of the new genetics in general and more specifically

(ii)    to allow the specialist, with little experience of genetics or its nomenclature, an entry into the world of genetic testing as it pertains to his/her specialty

These monographs are not intended as comprehensive accounts of each specialty — such reference texts are already available. Emphasis has been placed on those disorders with a strong genetic etiology and, in particular, those for which diagnostic testing is available.

The glossary is designed as a general introduction to molecular genetics and its language.

The revolution in genetics has been paralleled in recent years by the information revolution. The two complement each other, and the World Wide Web is a rich source of information about genetics. The following sites are highly recommended as sources of information:

1. PubMed. Free on-line database of medical literature. http://www.ncbi.nlm.nih.gov/PubMed/

2. NCBI. Main entry to genome databases and other information about the human genome project. http://www.ncbi.nlm.nih.gov/

3. OMIM. On line inheritance in Man. The On-line version of McKusick's catalogue of Mendelian Disorders. Excellent links to PubMed and other databases. http://www.ncbi.nlm.nih.gov/omim/

4. Mutation database, Cardiff. http://www.uwcm.ac.uk/uwcm/mg/hgmd0.html

Eli Hatchwell
Cold Spring Harbor Laboratory

# Preface

For the practicing endocrinologist, the wealth of information available from the past two decades of genetic research is truly a double-edged sword. The ability to pinpoint the genetic defects responsible for a specific endocrine disorder opens the possibility of faster and simpler diagnosis, improved understanding of disease mechanisms, and development of new treatment modalities. At the same time, the very abundance of information attained from the 'genetic revolution' may be so overwhelming that the practicing physician may fail to apply this knowledge to the daily routine of clinical practice.

Some clinicians, especially those whose training took place before the dawn of molecular genetics as we know it today, may find it difficult to understand the language and concepts of contemporary genetic research. Even if one is well versed in modern genetic laboratory procedures, few can be expected to stay abreast of new mutations, linkage associations, and other discoveries that are rapidly emerging from research laboratories. To a greater and greater degree, endocrinologists are being asked to provide genetic counseling, a role for which few have received formal training.

What patterns of inheritance are seen in familial growth hormone deficiency? What is the risk to the first-degree relatives of a child with type 1 diabetes mellitus? Is there a role for genetic studies in a patient with Noonan's syndrome? The answers to these and hundreds of other practical questions, that arise daily in the care of patients referred for the evaluation and treatment of endocrine disorders, can quickly and easily be found in *Genetics for Endocrinologists*. Before devoting hours to searching for the most current research articles, or spending time on the internet, we suggest you turn to *Genetics for Endocrinologists*. This concisely written volume is a valuable source of information in the busy clinic or at bedside rounds.

By referring to this book, physicians should be able to determine which patients and families should be referred to the genetics clinic for DNA studies, when appropriate, and for genetic counseling. Hopefully, *Genetics for Endocrinologists* will lead to closer contact between clinical endocrinologists and geneticists by clarifying the molecular basis and clinical expression of endocrine diseases.

Within this manual, we provide a concise summary of the most current information available on over 100 endocrine diseases. Where space constraints do not permit inclusion of the full spectrum of certain diseases, we present representative examples: hypochondroplasia is offered as a prototype for skeletal dysplasias, while Denys–Drash syndrome illustrates disorders of embryonic gonadal differentiation. Other conditions are included despite the fact that they are not, strictly speaking, hormonal diseases. For example, Prader–Willi, Marfan's, and Russell–Silver syndromes are often referred to an endocrinologist for diagnostic evaluation, and, in some cases, to co-ordinate treatment. Other nonendocrine conditions, such as thalassemia and hemochromatosis, have a very high incidence of associated endocrine complications.

In addition, diagrams of metabolic pathways are included to provide helpful summaries for medical students and endocrine fellows, as well as senior physicians. The glossary of genetic terminology can serve as an introduction to the language of genetics, and also act as a mini-reference dictionary.

Many of the diseases included in this manual show a wide range of phenotypic expression for any given mutation. The lack of a strong genotype–phenotype correlation may give rise to discontent with the value of the human genome project, or, alternatively, to a greater appreciation of the remarkable complexity of biological systems. Posttranslational modifications of proteins, environmental factors, and interactions with mutations or polymorphisms at other loci may explain in part why the same mutation can give rise to variable clinical features in different individuals. For example, in Bardet–Biedl syndrome, mutations occurring at two or more different loci are necessary to produce the clinical phenotype.

Every effort has been made to utilize the most current medical literature as the source for disease summaries in this volume. If the mini-review on a particular disease or syndrome stimulates the reader to learn more about this topic, *Genetics for Endocrinologists* can continue to help in two ways. Firstly, one can turn to the review articles and other citations from current medical journals listed at the end of each entry. Secondly, one can consult the OMIM website, using the MIM numbers listed for each specific disease. By linking this handbook to the OMIM site, one combines the advantages of easy, immediate access to a concise review of the current knowledge regarding a particular endocrine disorder, along with continuing updates, as new mutations and genetic information are published.

As public awareness of the human genome project continues to grow, and patients and their families have unlimited access to medical information via the internet, the clinical endocrinologist must be able to offer his or her patients the genetic information they demand and deserve to receive. It is our hope that *Genetics for Endocrinologists* will assist the busy clinician in providing answers to these increasingly complex and sophisticated questions.

# Contents

## 3. Disorders of the thyroid gland

## 4. Disorders of calcium and bone metabolism

# 5. Disorders of glucose metabolism

# 6. Genetic defects of the adrenal cortex, renal electrolyte balance, and the adrenal medulla

# 1. Multigland Syndromes

# Syndromes associated with endocrine gland tumors and hyperfunction

To begin with, four different tumor syndromes are described. The first, multiple endocrine neoplasia (MEN) type 1, is caused by dominant germline mutations in a tumor suppressor gene, *MEN1*. Tumors arise after a second somatic mutation occurs, affecting the normal *MEN1* allele.

The second two syndromes are caused by dominant mutations that activate the *RET* proto-oncogene. Mutations in this gene can cause at least four different syndromes, of which two, MEN2a and MEN2b, are described here. These are divided into separate entries because the clinical characteristics of the syndromes are quite different. A third endocrine syndrome caused by *RET* mutations, familial medullary thyroid carcinoma (MIM 155240), is described in Chapter 3.

The fourth entry, McCune–Albright syndrome, is caused by a *de novo* somatic mutation in a G-protein subunit, a critical factor in intracellular signal transduction.

# Syndromes associated with decreased endocrine function

Autoimmune polyendocrinopathy type 1, immunodysregulation, polyendocrinopathy, enteropathy, X-linked syndrome (IPEX), and polyglandular autoimmune syndrome type 2 are caused by autoimmune destruction of endocrine glands, whereas the remaining three syndromes are caused by damage to endocrine glands due to iron overload. Although the primary pathology is not endocrine, the final three syndromes are presented here because they have important endocrine manifestations, which are distinct because of differing etiologies of the iron overload.

# Multiple Endocrine Neoplasia Type 1

(also known as: MEN1; multiple endocrine adenomatosis type 1 [MEA1]; Wermer syndrome)

| | |
|---|---|
| **MIM** | 131100 |

| | |
|---|---|
| **Clinical features** | Primary hyperparathyroidism, almost always in the form of multiple parathyroid adenomas or hyperplasia, is the most common and usually the earliest manifestation of MEN1, occurring in 80%–100% of patients by the age of 40 years. About 50% of patients will develop gastrinomas. Nonfunctioning islet cell tumors or tumors that produce pancreatic polypeptide are common, frequently multiple, and often become malignant. Pituitary tumors are present in 10%–50% of MEN1 patients, prolactinomas and growth hormone-secreting tumors being the most prevalent. Foregut carcinoid tumors (bronchial, thymic, and gastric) are less common (10% of cases) and do not usually produce hormonal symptoms, but may become malignant. Skin manifestations, previously thought to be rare, in fact appear to be quite common. Specifically, facial angiofibromas have been reported in about 90% of patients and collagenomas in >70% of cases. About one third of patients have cutaneous or visceral lipomas. Pheochromocytomas have been reported, but are rare. Up to 40% of patients may have adrenal cortical tumors. Most are bilateral and nonfunctional, causing minimal morbidity; however, functional adrenal cortical tumors and adrenocortical carcinoma have been reported. Thyroid tumors have also been reported, but it is not clear if these are more common than in the general population. |

| | |
|---|---|
| **Primary tissue or gland affected** | Parathyroid, pancreatic islets, pituitary. |

| | |
|---|---|
| **Other organs, tissues, or glands affected** | Skin, adipose tissue (lipomas), adrenal cortex and medulla, thyroid. |

| | |
|---|---|
| **Age of onset** | Typically before 40 years of age. |

| | |
|---|---|
| **Epidemiology** | Prevalence estimates range from 0.01 to 2.5 per thousand population. |
| **Inheritance** | Autosomal dominant. |
| **Chromosomal location** | 11q13 |
| **Genes** | *MEN1* (menin). |
| **Mutational spectrum** | Over a hundred different mutations have been identified, including missense mutations, nonsense mutations, deletions, insertions, and splice-site mutations. No clear genotype–phenotype relationship has been identified. |
| **Effect of mutation** | Mutations interfere with menin's function as a repressor of JunD-mediated transcriptional activation. JunD is a transcription factor with antiapoptotic activity, which is thought to be mediated through inhibition of p53-dependent apoptosis. Lack of JunD suppression due to a lack of functional menin would result in decreased apoptosis, and thus have a net oncogenic effect. Patients carry one wild-type and one mutant *MEN1* allele, which is usually inherited, although it arises *de novo* in about 10% of patients. Tumors occur as described by Knudson's 'two hit' hypothesis. In a heterozygote individual, a somatic mutation in a single cell results in loss of the wild-type allele (loss of heterozygosity, LOH), leaving the cell with only the mutant gene. This cell proliferates, producing a monoclonal tumor. These somatic mutations occur randomly during life, explaining the late-onset of tumors. The typical age of onset and frequency of the various cell types affected is thought to reflect the propensity of each cell type to undergo somatic LOH. |
| **Diagnosis** | Diagnosis is based on finding an appropriate constellation of tumors, typically multiple parathyroid tumors, along with single or multiple islet cell or pituitary tumors. Documentation of autosomal dominant inheritance is particularly useful. Genetic diagnosis is definitive, but |

may be difficult because the entire sequence of the gene must be screened. Furthermore, in about 30% of cases, no mutation can be identified using currently available techniques.

**Counseling issues**    About 10% of *MEN1* germline mutations arise *de novo*. Furthermore, no mutation can be identified in approximately 10%–30% of MEN1 patients, suggesting that either the responsible mutations lie outside the gene regions tested, or that mutations in another gene may cause a clinically similar syndrome. Screening is important for early identification and treatment of tumors. Genetic studies will identify family members for whom screening can be stopped, and those who require intensive screening. However, since there is no specific intervention that can prevent or cure MEN1-associated cancers, knowledge of *MEN1* carrier status may not lead directly to any specific intervention.

**Notes**    Screening for tumors in MEN1 patients may help to reduce morbidity, though it rarely provides an opportunity for cure or prevention. The earliest reported presentation of MEN1 was at 5 years old, and 95% of patients have some disease penetrance by the age of 40 years. Annual screening for hypercalcemia, hyperprolactinemia, acromegaly, and gastrinoma should be performed. Imaging of the pituitary and the pancreas should be performed at baseline and every 3–5 years for life.

**References**    Schussheim DH, Skarulis MC, Agarwal SK et al. Multiple endocrine neoplasia type 1: New clinical and basic findings. *Trends Endocrinol Metab* 2001;12:173–8 (Review).

# Multiple Endocrine Neoplasia Type 2a

(also known as: MEN2a; multiple endocrine adenomatosis type 2a [MEA2a]; Sipple's syndrome)

| | |
|---|---|
| **MIM** | 171400 |
| **Clinical features** | Medullary thyroid carcinoma (MTC) is present in almost 100% of MEN2a patients, presenting either as a thyroid mass, or with diarrhea or flushing. MEN2a is found in approximately 15% of patients presenting with MTC, and is more common if the disease is bilateral or if diagnosis is before the age of 40 years. Pheochromocytoma is diagnosed in about 50% of MEN2a patients and may be bilateral. Hyperparathyroidism due to parathyroid hyperplasia is found in 25% of MEN2a patients. Most MEN2a patients present with symptoms related to MTC, though 10%–25% of cases may present with pheochromocytoma-related symptoms, the MTC only becoming evident after specific tests are performed. Hypercalcemia due to parathyroid hyperplasia is usually mild. Cutaneous lichen amyloidosis may be found. Rarely, patients may present with Hirschsprung's disease or megacolon. |
| **Primary tissue or gland affected** | Parathyroid, thyroid (C cells), adrenal medulla. |
| **Other organs, tissues, or glands affected** | Skin, colon. |
| **Age of onset** | The mean age of diagnosis is 27 years, with almost 100% of patients presenting by 40 years. |
| **Epidemiology** | Rare; accurate incidence or prevalence data are not available. |
| **Inheritance** | Autosomal dominant. |
| **Chromosomal location** | 10q11.2 |

| Genes | *RET* (ret tyrosine kinase). |
|---|---|
| **Mutational spectrum** | A total of 93%–98% of MEN2a patients have mutations in one of the five conserved cysteine residues in exons 10 and 11 of the *RET* proto-oncogene. Rarely, mutations in exons 13 and 14 have been reported. Hirschsprung's disease and megacolon are usually associated with mutations in *RET* codons 618 or 620, while cutaneous lichen amyloidosis is associated with a codon 634 mutation. The proband has a *de novo* mutation in about 5% of cases. |
| **Effect of mutation** | Mutations activate *RET* to function as a dominant oncogenic or transforming gene. |
| **Diagnosis** | Clinical diagnosis is based on finding the appropriate constellation of endocrine disease, particularly in a patient with a positive family history. MEN2a should be distinguished from familial (F)MTC and MEN2b, two autosomal dominant syndromes that are associated with specific mutations in the *RET* proto-oncogene. Clinical diagnosis of MTC is made by finding an elevated baseline serum calcitonin level, or an exaggerated calcitonin response to calcium and/or pentagastrin stimulation. Fine needle aspiration biopsy may be misleading unless specifically stained for calcitonin. Histology is necessary to confirm the diagnosis of MTC. Identification of bilateral MTC and/or C-cell hyperplasia greatly increases the risk of familial disease (MEN2a, MEN2b, or FMTC). Definitive diagnosis is made by identifying a germline mutation in the *RET* proto-oncogene. A mutation in codon 634 of exon 11 is found in 85% of patients. Sequencing of exons 10, 13, and 14 will identify mutations in another 13% of cases. No *RET* mutation can be identified in 2% of cases. |
| **Counseling issues** | All patients with MTC must be informed of the 25% chance of familial disease. If a mutation is identified, all family members should be screened. Siblings and children of the proband have a 50% chance of inheriting the mutation. Penetrance of disease |

approaches 100%. Family members carrying a mutation should be tested for MTC and its precursor, C-cell hyperplasia. Total thyroidectomy should be performed if calcitonin levels are abnormal or if there is an abnormal response to calcium–pentagastrin stimulation. Prophylactic total thyroidectomy should be considered, even if the calcitonin response is normal.

**Notes**

Early diagnosis is critical to long-term survival, since total thyroidectomy will cure or prevent malignant disease if it is performed before any extrathyroidal metastatic spread. Since metastatic MTC is frequently fatal, and carefully performed total thyroidectomy is associated with relatively low morbidity, prophylactic total thyroidectomy must be considered in all family members shown to have *RET* mutations, even if calcitonin levels are normal and no discrete tumor can be identified.

**References**

Modigliani E, Franc B, Niccoli-Sire P. Diagnosis and treatment of medullary thyroid cancer. *Baillieres Best Pract Res Clin Endocrinol Metab* 2000;14:631–49.

# Multiple Endocrine Neoplasia Type 2b

(also known as: MEN2b; multiple endocrine adenomatosis type 2b [MEA2b]; mucosal neuromata with endocrine tumors; mucosal neuroma syndrome; multiple endocrine neoplasia type 3 [MEN3]; Wagenmann–Froboese syndrome)

| | |
|---|---|
| **MIM** | 162300 |

**Clinical features**

The medullary thyroid carcinoma (MTC) associated with MEN2b is particularly aggressive, and begins in early childhood. Pheochromocytoma is found in 65% of cases, but it appears years after the diagnosis of MTC is made. Associated abnormalities typically include megacolon, intestinal ganglioneuromata, neuromata of the lips and tongue, and eyelid and corneal neuromata with medullated corneal nerve fibers. **Figure 1** shows a close-up view of the lips, tongue, and eyelid mucosa of a patient with MEN2b. Note the characteristic submucosal neurofibromas, which appear as light-colored, firm nodules that are typically found in the mouth and on the eyelid mucosa. Developmental and skeletal abnormalities are also common, and include marfanoid body habitus, coarse-appearing facies, thickened, anteverted eyelids, pectus excavatum, scoliosis, kyphosis, lordosis, and joint laxity.

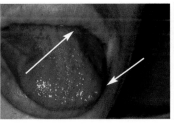

**Figure 1.** The lips, tongue, and eyelid mucosa of a patient with MEN2b. The white arrows indicate submucosal neurofibromas.

**Primary tissue or gland affected**

Thyroid (C cells), adrenal medulla.

| | |
|---|---|
| **Other organs, tissues, or glands affected** | Skin, mucus membranes, colon, cornea, bones and joints. |
| **Age of onset** | Childhood; C-cell hyperplasia has been identified in the first year of life. The mean age of diagnosis is 17 years, by which time metastatic disease is invariably present. |
| **Epidemiology** | Very rare. About 50% of cases are caused by *de novo* mutations. |
| **Inheritance** | Autosomal dominant. |
| **Chromosomal location** | 10q11.2 |
| **Genes** | *RET* (ret tyrosine kinase). |
| **Mutational spectrum** | A single mutation in codon 918, exon 16 of the *RET* gene causes 95% of cases. Rarely, mutations in exons 15 and 16 have been reported. |
| **Effect of mutation** | Mutations activate *RET* to function as a dominant oncogenic or transforming gene. |
| **Diagnosis** | The clinical diagnosis is made by identifying MTC associated with other cutaneous or skeletal abnormalities. Definitive diagnosis is made by identifying a mutation in the *RET* gene. The vast majority of patients have mutations in codon 918 of *RET*, making genetic diagnosis simple. A minority of patients have other *RET* mutations, while no mutation can be identified in 4%–5% of patients with the clinical syndrome. |
| **Counseling issues** | The disease is inherited as an autosomal dominant trait with very high penetrance. Children of affected individuals have a 50% chance of inheriting the disease. Unless the proband has a *de novo* mutation, siblings also have a 50% chance of carrying the disease. Malignant MTC is the cause of death in most patients; early diagnosis is important, and total thyroidectomy before the |

appearance of clinical disease is the only curative treatment. Therefore, the siblings and children of affected individuals should be screened as soon after birth as possible. Total thyroidectomy should be performed as soon as is technically possible, preferably within the first year of life. Because of the potentially lethal consequences of untreated pheochromocytoma, life-long screening is mandatory in patients with the molecular diagnosis of MEN2b.

**References**

Modigliani E, Franc B, Niccoli–sire P. Diagnosis and treatment of medullary thyroid cancer. *Baillieres Best Pract Res Clin Endocrinol Metab* 2000;14:631–49.

# McCune–Albright Syndrome

(also known as: MAS; polyostotic fibrous dysplasia with café-au-lait pigmentation)

| | |
|---|---|
| **MIM** | 174800 |
| **Clinical features** | Solitary or multiple skeletal fibrous dysplasia lesions are most commonly found in the femur and pelvis, and cause progressive deformity, fractures, and nerve entrapment. Café-au-lait skin lesions with irregular borders ('coast of Maine') are most often seen on the lower back and buttocks, usually on the same side as the skeletal lesions, and, in contrast to lesions of neurofibromatosis, rarely cross the midline. Endocrine dysfunction is characterized by autonomous hypersecretion of the affected endocrine glands. Gonadotropin-independent precocious puberty is common. Ovarian estrogen production may vary as ovarian cysts grow and involute. Precocious puberty is less common in males and is characterized by Leydig cell hyperplasia. Gonadotropin levels are typically low or suppressed. MAS patients with excess growth hormone secretion resemble those with acromegaly or gigantism due to spontaneous tumors. Hyperprolactinemia is present in most MAS patients with growth hormone excess. Hyperthyroidism and autonomous functioning thyroid nodules may be found in 33% of MAS patients. Cushing's syndrome in MAS patients is due to nodular hyperplasia or a solitary adenoma of the adrenal cortex. Hypophosphatemic rickets may be seen in patients with polyostotic fibrous dysplasia, even in the absence of other features of MAS. Other nonendocrine manifestations include cardiac arrhythmias, congestive heart failure, and severe neonatal cholestasis. |
| **Primary tissue or gland affected** | Ovaries, testes, thyroid, adrenal cortex, anterior pituitary. |
| **Other organs, tissues, or glands affected** | Bone, skin. |

| | |
|---|---|
| **Age of onset** | Endocrine manifestations such as precocious puberty or Cushing's syndrome may appear before the age of 5 years. Skeletal lesions appear within the first 10 years of life. |
| **Epidemiology** | A Medline search from 1966 through to 1996 identified 158 reported cases of MAS. |
| **Inheritance** | Sporadic. |
| **Chromosomal location** | 20q13.2 |
| **Genes** | *GNAS1* (stimulatory $\alpha$ subunit of G protein). |
| **Mutational spectrum** | A point mutation at Arg201 has been found in all affected subjects. The mosaic distribution of cells containing mutant *GNAS1* varies according to the degree to which each tissue is affected. Typically, the mutation occurs early in embryogenesis; mutations arising later in embryogenesis result in fewer affected cells and a milder phenotype as compared with earlier mutational events. |
| **Effect of mutation** | Mutations of the *GNAS1* gene at Arg201 inhibit GTPase activity, leading to constitutive activation of adenylyl cyclase. This, in turn, causes increased production of cAMP and activation of protein kinase A in affected cells, resulting in cellular proliferation and hormone secretion (see **Figure 2**). |
| **Diagnosis** | Autonomous hypersecretion of affected endocrine glands results in low or suppressed levels of tropic hormones (adrenocorticotropic hormone, thyrotropin, luteinizing hormone [LH], and follicle-stimulating hormone [FSH]). Although gonadotropin levels are typically suppressed in girls with MAS and precocious puberty, gonadotropin-releasing hormone stimulation may elicit a normal LH and FSH response during intervals of ovarian inactivity, and in some girls who develop 'true' central precocious puberty. In contrast to the 'usual' patient with autoimmune thyroid disease, in MAS |

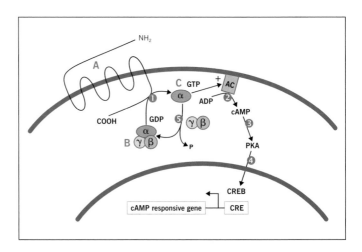

**Figure 2.** Schematic representation of the mechanism by which Gs-α-1 mutations result in constitutive transcription of cAMP-responsive genes. Stimulation of a membrane-bound G protein-coupled receptor (**A**) results in exchange of GDP with GTP (reaction **❶**) on the α subunit of the G protein heterotrimer (**B**). This causes dissociation and activation of the α subunit (**C**), which in turn stimulates adenylyl cyclase (AC) to form cAMP (reaction **❷**). Elevated cAMP cause activation of protein kinase A (PKA) (reaction **❸**), which in turn phosphorylates and activates the cAMP-responsive element binding protein (CREB) which translocates to the nucleus (reaction **❹**), binds CRE and activates transcription of cAMP-responsive genes. Under normal circumstances, activation of this cascade is stopped by the endogenous GTPase activity of the α subunit (reaction **❺**). Arg201 mutations in the α subunit abolish this GTPase activity, resulting in accumulation of activated α subunit and constitutive activation of adenylyl cyclase.

patients with hyperthyroidism or thyroid nodules, antithyroid antibodies and thyroid-stimulating immunoglobulins are negative.

**Counseling issues**

Families of affected patients should be reassured that MAS is a sporadic condition. A thorough evaluation is necessary in order to search for the variable manifestations of the syndrome.

**Notes**

Polyostotic fibrous dysplasia may occur as isolated lesions without other manifestations of MAS. Management of the skeletal lesions is a major therapeutic challenge. Bisphosphonates, such as pamidronate, have achieved some degree of success. Precocious puberty is treated with aromatase inhibitors, including testolactone, ketaconazole, and, more recently, anastrozole. Octreotide and bromocriptine are effective in some MAS patients with acromegaly and hyperprolactinemia. Thyroid nodules may be treated with radioactive iodine or surgery.

**References**

Lania A, Mantovani G, Spada A. G protein mutations in endocrine diseases. *Eur J Endocrinol* 2001;145:543–59 (Review).

Levine MA. Clinical implications of genetic defects in G proteins: Oncogenic mutations in G alpha s as the molecular basis for the McCune–Albright syndrome. *Arch Med Res* 1999;30:522–31 (Review).

# Autoimmune Polyendocrinopathy Type 1

(also known as: autoimmune polyendocrinopathy–candidiasis–ectodermal dystrophy [APECED]; polyglandular autoimmune syndrome type 1 [PGA1]; polyglandular deficiency syndrome [Persian-Jewish type])

| | |
|---|---|
| **MIM** | 240300 |
| **Clinical features** | Candidiasis is usually the first symptom. Endocrinopathies develop later, with hypoparathyroidism usually preceding adrenal insufficiency. Hypogonadism is also common, but type 1 diabetes mellitus and autoimmune thyroid disease are infrequent. Hypogammaglobulinemia, chronic active hepatitis, pernicious anemia, alopecia, and keratopathy may also be present. |
| **Primary tissue or gland affected** | Parathyroid, adrenal, thyroid, gonads, pancreatic β cells. |
| **Other organs, tissues, or glands affected** | Skin and mucus membranes (candidiasis), liver, bone marrow, hair. |
| **Age of onset** | The disease usually becomes evident during the first two decades of life. |
| **Epidemiology** | The overall worldwide incidence is very low; however, the syndrome is more commonly seen in specific, genetically isolated populations. An incidence of 1:6500–9000 has been reported in Persian Jews. A similar incidence is found in some regions of Finland. An increased incidence in the Sardinian population has also been reported. |
| **Inheritance** | Autosomal recessive. |
| **Chromosomal location** | 21q22.3 |
| **Genes** | *AIRE* (autoimmune regulator). |

**Mutational spectrum**      A single missense mutation (R257X) is the predominant mutation in Finnish APECED patients. Missense, nonsense, and frame-shift mutations have been identified in other populations.

**Effect of mutation**      The mutations result in loss of expression or loss of function of the protein.

**Diagnosis**      The diagnosis should be suspected in a child with unexplained candidiasis, particularly if the child is of Finnish or Persian-Jewish extraction. Clinical diagnosis is based on finding the appropriate constellation of autoimmune diseases. Genetic diagnosis is possible and is definitive.

**Counseling issues**      Awareness of genetic predilection to this entity will aid early diagnosis, which can significantly reduce morbidity from the various hormone deficiencies.

**Notes**      Some phenotypic differences in the syndrome have been described between those of Finnish and Persian-Jewish decent. These are probably caused by allelic heterogeneity.

**References**      Nagamine K, Peterson P, Scott HS et al. Positional cloning of the APECED gene. *Nat Genet* 1997;17:393–8.

Zlotogora J, Shapiro MS. Polyglandular autoimmune syndrome type I among Iranian Jews. *J Med Genet* 1992;29:824–6.

# Immunodysregulation, Polyendocrinopathy, Enteropathy, X-linked Syndrome

(also known as: IPEX; X-linked autoimmunity–allergic dysregulation syndrome [XLAAD]; insulin-dependent diabetes mellitus–secretory diarrhea syndrome [DMSD]; X-linked polyendocrinopathy, immune dysfunction and diarrhea [XPID]; absence of islets of Langerhans)

| | |
|---|---|
| **MIM** | 304790, 304930 |
| **Clinical features** | Patients present with neonatal, brittle insulin-dependent diabetes, severe diarrhea, eczema, hemolytic anemia, thyroiditis, and intrauterine growth retardation. Immunoglobulin E levels are high. Immune dysregulation results in susceptibility to severe infections, including sepsis and fungal infections. Small bowel biopsy shows severe villous atrophy and mucosal erosion with lymphocytic infiltration. Anti-islet cell and antithyroid antibodies are present in some, but not all, patients. Survival beyond the age of 2 years is rare, with most patients succumbing to infections within the first year of life. |
| **Primary tissue or gland affected** | Pancreatic β cells. |
| **Other organs, tissues, or glands affected** | Immune system, gastrointestinal tract, skin, thyroid. |
| **Age of onset** | Birth. |
| **Epidemiology** | This is a very rare syndrome, the incidence of which has not been studied. |
| **Inheritance** | X-linked recessive. |
| **Chromosomal location** | Xp11.23-q13.3 |
| **Genes** | *FOXP3* (forkhead box P3). |

| | |
|---|---|
| **Mutational spectrum** | Nine different missense, splice, and deletion mutations have been identified. In one family, a rare polyadenylation signal mutation cosegregated with disease. |
| **Effect of mutation** | Mutations appear to result in failure to produce a functional protein. |
| **Diagnosis** | Clinical diagnosis is based on the constellation of defects that characterize the syndrome. Identification of *FOXP3* mutations will confirm the diagnosis, though it is not known if this gene is involved in the pathogenesis of disease in all patients with this syndrome. |
| **Notes** | The enteropathy is autoimmune mediated, and treatment with cyclosporin A, FK506, and steroids appears to be beneficial. Bone marrow transplantation has been reported in two patients and appeared to result in a marked amelioration of symptoms. Interestingly, the immunosuppressive treatment given to prevent rejection of the allograft appeared to have independent marked beneficial effects. Insulin-dependent diabetes resolved in one patient, but persisted in the other. The scurfy syndrome in mice is phenotypically similar to IPEX and is caused by mutations in the murine equivalent of the human *FOXP3* gene. This mouse model may be used to better understand the pathophysiology of the syndrome, and to develop novel treatment modalities. |
| **References** | Bennett CL, Ochs HD. IPEX is a unique X-linked syndrome characterized by immune dysfunction, polyendocrinopathy, enteropathy, and a variety of autoimmune phenomena. *Curr Opin Pediatr* 2001;13:533–8 (Review). |
| | Bennett CL, Christie J, Ramsdell F et al. The immune dysregulation, polyendocrinopathy, enteropathy, X-linked syndrome (IPEX) is caused by mutations of FOXP3. *Nat Genet* 2001;27:20–1. |
| | Wildin RS, Ramsdell F, Peake J et al. X-linked neonatal diabetes mellitus, enteropathy and endocrinopathy syndrome is the human equivalent of mouse scurfy. *Nat Genet* 2001;27:18–20. |

# Polyglandular Autoimmune Syndrome Type 2

**(also known as: Schmidt's syndrome; autoimmune polyendocrine syndrome type 2)**

| | |
|---|---|
| **MIM** | 269200 |
| **Clinical features** | Clinical features include several endocrine and nonendocrine autoimmune defects, including autoimmune thyroid disease (hyperthyroidism or hypothyroidism); Addison's disease; type 1 diabetes mellitus; hypoparathyroidism; pernicious anemia; chronic candidiasis of mucosa, skin, and nails; vitiligo; chronic hepatitis; cirrhosis; and alopecia. T-lymphocyte deficiency, splenic agenesis, and thymic dysplasia may also be present. |
| **Primary tissue or gland affected** | Thyroid, pancreatic $\beta$ cells, adrenal cortex. |
| **Other organs, tissues, or glands affected** | Hematopoietic system, mucus membranes (candidiasis), skin, nails, hair, liver, spleen, thymus. |
| **Age of onset** | Patients usually present during the second and third decades of life, with a mean age of onset of 24 years. |
| **Epidemiology** | Unknown. |
| **Inheritance** | Unknown. |
| **Chromosomal location** | Unknown. |
| **Genes** | Unknown. |
| **Mutational spectrum** | In some families, linkage to the human leukocyte antigen (HLA) locus on chromosome 6 suggests a primary immunologic derangement. However, linkage to this region has been excluded in other families. |

| | |
|---|---|
| **Effect of mutation** | Unknown. |
| **Diagnosis** | This syndrome should be suspected when autoimmune thyroid disease (either hyperthyroidism or hypothyroidism) is associated with Addison's disease. Antibodies to 21-hydroxylase can be found in most patients with Addison's disease. |
| **Counseling issues** | Since the genetics of the syndrome are not yet understood, no specific recommendations can be made. A high index of suspicion in family members is important to ensure early diagnosis of endocrine and immune deficiencies. |
| **Notes** | Clinical heterogeneity suggests that this may be a genetically heterogeneous syndrome. Identification of specific genetic abnormalities in multigenerational families and genetic isolates may help to elucidate the precise etiology of this syndrome. Autosomal dominant inheritance is suspected, and multifactorial inheritance is also possible. Association with a specific HLA haplotype in one family suggests that the gene responsible may be in linkage disequilibrium with the HLA complex on chromosome 6. |
| **References** | Betterle C, Volpato M, Greggio AN et al. Type 2 polyglandular autoimmune disease (Schmidt's syndrome). *J Pediatr Endocrinol Metab* 1996;9(Suppl. 1):S113–23 (Review). |
| | Takeda R, Takayama Y, Tagawa S et al. Schmidt's syndrome: Autoimmune polyglandular disease of the adrenal and thyroid glands. *Isr Med Assoc J* 1999;1:285–6. |

# Hemochromatosis Type 1

**(also known as: HFE; hereditary hemochromatosis [HH or HHC])**

| | |
|---|---|
| MIM | 235200 |

**Clinical features**  Classical clinical features include hepatosplenomegaly leading to cirrhosis and eventually hepatocellular carcinoma, cardiomyopathy, skin hyperpigmentation, diabetes ('bronze diabetes'), joint disease, hypogonadotropic hypogonadism, and osteopenia. The latter is presumably related to decreased sex hormones secondary to hypogonadism. Coexistent alcohol abuse increases the risk of cirrhosis and hepatocellular carcinoma. Patients only rarely present with a classical combination of symptoms that makes the diagnosis obvious. More typically, patients present with unexplained hypogonadism, cardiomyopathy, cirrhosis, or hepatocellular carcinoma. Hyperpigmentation may be absent or so chronic and slowly progressive that the patient or their family fail to identify it as pathologic.

**Primary tissue or gland affected**  Pituitary gonadotrophs, pancreatic $\beta$ cells, thyroid.

**Other organs, tissues, or glands affected**  Skin, liver, heart, joints, bones.

**Age of onset**  The fourth or fifth decade of life.

**Epidemiology**  About 10% of white people from North America and northern Europe are heterozygous for the C282Y mutation on the *HFE* gene. Other ethnic groups have lower allele frequencies, suggesting a Celtic origin of this mutation. Specifically, an allele frequency of only 0.01 has been found in the Italian population, and 0.013 in the Ashkenazi Jews. Allele frequency is essentially zero in those from Africa and the Indian subcontinent. In some, but not all, studies, serum iron, transferrin, and ferritin levels were found to be higher in heterozygotes than in controls, but the clinical significance of this is not known. The H63D mutation on the same gene is more common worldwide (allele frequency 0.22 in European and North American populations), but carries a much lower risk of disease. Therefore, though in the general population H63D homozygotes are more common than C282Y homozygotes, the former frequently have no evidence of iron overload. In disease populations, the large majority of patients have the C282Y mutation.

**Inheritance**  Autosomal recessive.

**Chromosomal location**  6p21.3

**Genes**  *HFE* (hemochromotosis gene).

**Mutational spectrum**  As of 2001, 37 allelic variants had been described, two of which are particularly common. About 85% of *HFE* chromosomes have the C282Y mutation, and most of the others have the H63D mutation. Other mutations were mostly identified in patients heterozygous for the C282Y or H63D mutations.

**Effect of mutation**  Different mutations affect HFE function differently. The C282Y mutation abrogates binding of HFE to β2-microglobulin. This mutation results in a relatively severe functional defect, and homozygosity carries a very high risk of clinical disease (odds ratio 4,382). The H63D mutation appears to affect the affinity of HFE for the transferrin receptor (TFR). The functional significance of this defect is much less than for the C282Y mutation, and the penetrance of clinical disease is low (odds ratio 5.7), with most individuals who are homozygous for the H63D mutation showing no evidence of disease.

**Diagnosis**  HFE should be suspected when any combination of the above-described symptoms is found. However, only a minority of patients present with a classical constellation of symptoms, including hyperpigmentation, hepatosplenomegaly, and diabetes. Most will present with one or two of the typical findings, making diagnosis difficult unless the clinician has a high index of suspicion. Hemochromatosis should be suspected in any patient with unexplained hypogonadotropic hypogonadism, cardiomyopathy, or hepatocellular carcinoma. Since biochemical screening tests are easy to perform and readily available, and treatment significantly improves prognosis, an effort should be made to establish the diagnosis and initiate specific treatment as early as possible. Transferrin saturation, serum iron, and serum ferritin levels are typically elevated, with transferrin saturation being the most sensitive of the three tests. Liver biopsy with determination of the total iron content will confirm the diagnosis of iron overload, and genetic testing will confirm the specific etiology.

**Counseling issues**

Penetrance is variable, but is higher in males than in females since menstrual bleeding provides a natural mechanism for removing excess iron. Early diagnosis and treatment prevent complications and are therefore life-saving. Treatment is based on the clearance of excess iron load. This is most commonly performed by prophylactic phlebotomy, though chelating agents can also be used. Once an index case is identified, efforts should be made to identify related cases by family screening.

**Notes**

See also juvenile hemochromatosis (next entry), which is a more severe version of the disease and is caused by mutations in a different gene. In some patients with HFE, no mutation is found in the *HFE* gene. Mutations in the *TFR2* gene on chromosome 7q22 have been found in two such families. In addition, neonatal HFE, an African iron overload syndrome, and an autosomal dominant variant of HFE, have been described. The genetic etiologies of these are not understood.

**References**

Hanson EH, Imperatore G, Burke W. HFE gene and hereditary hemochromatosis: A HuGE review. Human Genome Epidemiology. *Am J Epidemiol* 2001;154:193–206 (Review).

Moczulski DK, Grzeszczak W, Gawlik B. Role of hemochromatosis C282Y and H63D mutations in HFE gene in development of type 2 diabetes and diabetic nephropathy. *Diabetes Care* 2001;24:1187–91.

Papanikolaou G, Politou M, Terpos E et al. Hereditary hemochromatosis: HFE mutation analysis in Greeks reveals genetic heterogeneity. *Blood Cells Mol Dis* 2000;26:163–8.

# Hemochromatosis Type 2

(also known as: HFE2; juvenile hemochromatosis [JH])

| | |
|---|---|
| **MIM** | 602390 |
| **Clinical features** | Abdominal pain is typically found during the first decade of life, hypogonadotropic hypogonadism during the second decade, and cardiac symptoms in the third. Joint symptoms may be prominent. If iron overload is not adequately treated, HFE2 is lethal because of cardiac complications that include arrhythmias and intractable heart failure. Cardiac symptoms may be prevented or improved with phlebotomy and/or chelation therapy. Pancreatic atrophy with or without diabetes may be present. In contrast to HFE1 (MIM 235200, see previous entry), males and females are equally affected. |
| **Primary tissue or gland affected** | Pituitary gonadotrophs, pancreatic $\beta$ cells, thyroid. |
| **Other organs, tissues, or glands affected** | Heart, joints, pancreatic exocrine tissue. |
| **Age of onset** | Clinical symptoms usually appear before the age of 30 years. |
| **Epidemiology** | The disease is rare. Consanguinity is frequently reported. Several apparently unrelated Italian families with the disease have been described. |
| **Inheritance** | Autosomal recessive. |
| **Chromosomal location** | 1q |
| **Genes** | Unknown. |

| | |
|---|---|
| **Mutational spectrum** | Unknown. |
| **Effect of mutation** | Unknown. |
| **Diagnosis** | The diagnosis of HFE is suspected on the basis of clinical symptoms. Elevated serum parameters of iron metabolism (which include serum iron and ferritin levels, and transferrin saturation) provide important additional information, but false-positive and false-negative results have been reported for all tests. Thus, the definitive diagnosis of iron overload is made by liver biopsy. HFE2 is suspected on clinical grounds, primarily clinical severity and age of onset of symptoms. Genetic testing for *HFE* mutations may be indicated to differentiate between HFE1 and HFE2 in some cases. |
| **Counseling issues** | Autosomal recessive inheritance with equal male to female distribution is expected. *HFE* mutation analysis or major histocompatibility complex class 1 linkage analysis is not helpful. Early diagnosis and treatment may prevent severe cardiac symptoms. |
| **Notes** | The few patients thoroughly studied have shown extremely avid absorption of radioactive iron, suggesting near complete loss of the normal regulatory mechanism responsible for maintaining iron balance. Heterozygosity for the common *HFE* mutations may be associated with more severe disease in some families. |
| **References** | Roetto A, Totaro A, Cazzola M et al. Juvenile hemochromatosis locus maps to chromosome 1q. *Am J Hum Genet* 1999;64:1388–93. |

# β-Thalassemias

(includes: thalassemia major, intermedia, and minor)

**MIM**             141900

**Clinical features**   The primary clinical feature is anemia. In thalassemia major,
the anemia is severe and requires regular transfusion therapy.
Thalassemia minor is asymptomatic, and is only identified when
blood tests are performed for other reasons. Thalassemia intermedia,
as its name implies, is of intermediate severity, usually requiring
transfusions only rarely or at times of clinical decompensation.
Recurrent transfusions result in severe iron overload. Since currently
available chelation therapy is of limited efficacy and quite
burdensome for the patient, most patients will have very marked
iron overload-related organ dysfunction despite chelation therapy.
The effect of iron overload on the endocrine system is very cell-type
specific. In recent reviews of over 1000 cases, diabetes was present
in 4.9%–5.4% of those who reached puberty, primary
hypothyroidism in 6.2%–11.6%, and hypogonadotropic
hypogonadism in 55%. Hypoparathyroidism is present in 3.6% of
patients, typically appearing in the third decade of life. Short stature
is associated with a relatively normal growth hormone (GH) reserve,
but abnormally low insulin-like growth factor (IGF)-1 levels. IGF-1
levels and growth rate may improve with GH replacement therapy,
suggesting a relative resistance to GH with an inadequate pituitary
response. Osteoporosis and bone deformities are common and are
probably related to the combination of hypogonadism,
hypoparathyroidism, and bone marrow expansion due to increased
hematopoiesis. Early diagnosis and adequate treatment improve the
quality of life of these patients. The thyroid is usually the first gland to
be clinically affected. Gonadotropins appear to be particularly
sensitive to iron-induced damage, but this only becomes clinically
relevant at puberty. Insulin-requiring diabetes is often a relatively late
occurrence and is associated with a very poor prognosis, primarily
because of concomitant iron-related cardiac disease.

| | |
|---|---|
| **Primary tissue or gland affected** | Thyroid, pituitary, parathyroid, pancreatic β cells. |
| **Other organs, tissues, or glands affected** | Hematopoietic system, heart, liver, bones. |
| **Age of onset** | During the first year of life. |
| **Epidemiology** | The incidence of thalassemia major is extremely variable, with carrier rates (thalassemia minor) of >20% in some countries of the Mediterranean basin. An incidence of 1:2700 has been recently reported in India. The disease is rare in populations originating from northern Europe and the Americas. |
| **Inheritance** | Autosomal recessive. |
| **Chromosomal location** | 11p15.5 |
| **Genes** | *HBB* (hemoglobin β-subunit). |
| **Mutational spectrum** | Several hundred different mutations have been identified. These include premature truncation due to a stop codon or frame shift, missense mutations, and deletions. Large deletions occur, but are uncommon. |
| **Effect of mutation** | The mutations result in a marked decrease in β-globulin production or in the production of nonfunctional protein. |
| **Diagnosis** | The diagnosis of β-thalassemia is based on hematologic findings, such as severe anemia and abnormal red cell morphology, and typical findings on bone marrow aspiration and biopsy. Bone marrow findings include marked erythroid hyperplasia with a high degree of immaturity and bizarre morphology of the erythroid progenitors. Early erythroblasts are particularly abundant (shift to left) and may appear megaloblastic. Alpha-globin inclusions are readily apparent if |

appropriate dyes are used. Endocrine deficiencies present as the child grows and the severity of iron overload increases. Routine screening of thyroid function, and calcium, sex hormone, gonadotropin, and glucose levels is recommended. When abnormalities in screening tests appear, a definitive diagnosis can be made for each endocrine axis according to standard techniques.

**Counseling issues**

Endocrine abnormalities in thalassemia major are relatively easily treated and are secondary to the hematologic abnormality and its treatment. Genetic counseling issues relate primarily to the hematologic aspects of the disease. Improved genetic diagnosis and modern technology can be utilized to considerably decrease the incidence of this disease.

**References**

Borgna-Pignatti C, Rugolotto S, De Stefano P et al. Survival and disease complications in thalassemia major. *Ann NY Acad Sci* 1998;850:227–31.

Italian Working Group on Endocrine Complications in Non-endocrine Diseases. Multicentre study on prevalence of endocrine complications in thalassaemia major. *Clin Endocrinol (Oxf)* 1995;42:581–6.

# 2. Pituitary Disease and Growth Retardation

# Syndromes associated with multiple pituitary hormone deficiency

Pituitary hormone deficiency can be caused by structural or developmental abnormalities of the pituitary, or by insensitivity to pituitary hormone action. In this chapter, we discuss several syndromes of multiple pituitary hormone deficiency. The first appears to be isolated to the pituitary itself, whereas the following three syndromes are associated with nonpituitary signs and symptoms, some of which may dominate the clinical picture.

# Syndromes associated with a single pituitary hormone deficiency

The next four entries discuss loss of a single pituitary hormone function. This can be caused by defects at several levels, including hypothalamic-releasing hormone, releasing hormone receptor, pituitary hormone itself, or the pituitary hormone receptor. In the case of growth hormone (GH) deficiency, entries include descriptions of mutations in the GH-releasing hormone receptor, GH itself, the GH receptor, and insulin-like growth factor-1, an intermediate needed for GH to exert most of its physiologic effects. Similarly, hypothyroidism can be caused by defects at the level of the hypothalamus, the pituitary, the thyroid itself, or peripheral thyroid hormone action. The entry on isolated central hypothyroidism discusses defects in the thyrotropin gene and its receptor. Defects at the level of the thyroid gland and thyroid hormone receptor are discussed in Chapter 3.

Central diabetes insipidus (cDI) is also discussed in this chapter. In contrast to the central defect, nephrogenous diabetes insipidus (nDI) is caused by defects in either of two genes expressed in the distal nephron. Although both cDI and nDI may enter into the differential diagnosis of a patient presenting with unexplained polyuria, the latter is discussed in Chapter 6, along with other disorders of renal fluid and electrolyte handling.

# Complex, nonendocrine syndromes

The chapter ends with a discussion of seven complex syndromes that are not strictly endocrine in pathophysiology. However, these syndromes frequently present as abnormal growth, and thus the pediatrician or pediatric endocrinologist is likely to be the first to be presented with the clinical problem.

# Multiple Pituitary Hormone Deficiencies

(also known as: MPHD; congenital hypopituitarism; combined pituitary hormone deficiencies; pituitary-specific transcription factor 1 [PIT1] deficiency; POU domain, class 1, transcription factor 1 [POU1F1] deficiency; prophet of PIT1 [PROP1] deficiency)

| | |
|---|---|
| **MIM** | 173110, 601538 |
| **Clinical features** | Typical features include severe growth retardation from infancy, a prominent forehead, midfacial hypoplasia with a depressed nasal bridge, and a short nose with anteverted nostrils. Some patients are detected on neonatal screening for congenital hypothyroidism. Later in childhood, growth retardation may be associated with mental retardation. Luteinizing hormone (LH) and follicle-stimulating hormone (FSH) production are unaffected in patients with PIT1 mutations; however, patients with PROP1 defects have hypogonadotropic hypogonadism and do not undergo spontaneous puberty. The degree and severity of symptoms and age at diagnosis for PROP1 defects are variable, even among patients with the same mutation. For both PIT1 and PROP1 mutations, pituitary size is small compared with age-matched controls, and temporary visualization of parasellar tissue masses that tend to regress over time has been noted in some patients with PROP1 defects. |
| **Primary tissue or gland affected** | Pituitary. |
| **Other organs, tissues, or glands affected** | Thyroid, adrenals, gonads (secondary dysfunction due to pituitary deficiency). |
| **Age of onset** | Infancy or childhood. |
| **Epidemiology** | A specific etiology is recognized in only 10%–20% of all patients treated for pituitary hormone deficiency. The percentage of hypopituitary patients with a similarly affected first-degree relative may be as high as 30%. The Arg271Trp dominant mutation appears |

to account for the majority of cases of PIT1-related MPHD. Among patients with MPHD, the overall frequency of *PROP1* mutations is much greater than that of *PIT1* mutations. A significant number of MPHD patients have neither identifiable *PIT1* or *PROP1* mutations.

**Inheritance**  Autosomal recessive, autosomal dominant.

**Chromosomal location**  3p11 (PIT1), 5q35 (PROP1).

**Genes**  *PIT1* (pituitary-specific transcription factor 1; also known as *POU1F1* [POU domain, class 1, transcription factor 1]), *PROP1* (prophet of PIT1).

**Mutational spectrum**  Recessive *PIT1* mutations include missense, nonsense, and frame-shift mutations, and deletion of the entire coding sequence. Dominant *PIT1* mutations result from amino acid substitutions. All *PROP1* defects described so far are recessive. These include frame-shift, missense, nonsense, and splice-error mutations resulting in loss of function.

**Effect of mutation**  PIT1 regulates the proliferation and differentiation of thyrotropin (TSH)-, growth hormone (GH)-, and prolactin-producing cells in the embryonic pituitary. In addition, PIT1 regulates GH synthesis by direct binding to the GH gene promoter. *PROP1* consists of three exons encoding a 233-amino acid protein. The PROP1 gene product precedes the appearance of PIT1 in the developing pituitary and is a prerequisite for PIT1 expression. Recessive *PIT1* mutations may lead to complete absence of protein production, synthesis of a truncated protein, decreased DNA binding, or normal DNA binding, but an inability to stimulate transcription. A dominant mutation in which tryptophan is substituted for arginine at position 271 results in mutant protein with an increased affinity for GH and prolactin promoter sites. The mutant proteins preferentially occupy the binding sites and block, rather than stimulate, transcription. In contrast to *PIT1*, the influence of *PROP1* is limited to a specific time-frame of fetal development. Pituitary cells that 'escape' the adverse effects of a *PROP1* mutation during fetal development may function without adverse effects from the mutation in postnatal life. As a consequence, some residual hormonal function is observed in patients with *PROP1* mutations, while GH and prolactin levels are typically undetectable in those with *PIT1* defects.

**Diagnosis**

Both *PIT1* and *PROP1* mutations need to be considered as possible causes of familial MPHD. Secondary hypothyroidism along with low or undetectable levels of prolactin, lack of TSH and prolactin response to thyrotropin-releasing hormone stimulation, and an inadequate response of GH to standard stimulation tests point to a defect in *PIT1* or *PROP1*. LH and FSH levels are unaffected by *PIT1* mutations, but are low in patients with *PROP1* defects. Cortisol levels are usually normal, though secondary adrenal insufficiency has been noted in approximately one third of patients with *PROP1* defects. Organic causes of MPHD, mainly craniopharyngioma and other intracranial tumors, need to be excluded. *PROP1* mutations can present with pituitary enlargement in early childhood. *PIT1* and *PROP1* mutations have not been found in MPHD patients with a hypoplastic pituitary stalk or ectopic posterior pituitary on magnetic resonance imaging.

**Counseling issues**

Families need to be advised of the possibility of dominant as well as recessive modalities of inheritance. Thyroxine levels should be measured in all subsequent infants in order to diagnose and treat congenital hypothyroidism as early as possible. Diagnosis of a *PROP1* mutation in a prepubertal child will allow for timely initiation of treatment for hypogonadism.

**Notes**

Early recognition of congenital hypothyroidism is critical in order to prevent irreversible neurologic damage in affected children. Neonatal screening programs that only measure TSH levels will not detect infants with secondary hypothyroidism. Some patients with MPHD have normal *PIT1* and *PROP1* coding sequences.

**References**

Parks JS, Brown MR, Hurley DL et al. Heritable disorders of pituitary development. *J Clin Endocrinol Metab* 1999;84:4362–70 (Review).

Pfaffle RW, Blankenstein O, Wuller S et al. Combined pituitary hormone deficiency: Role of Pit-1 and Prop-1. *Acta Paediatr Suppl* 1999;88:33–41 (Review).

# Hypopituitarism and Rigid Cervical Spine Syndrome

(also known as: combined pituitary hormone deficiency due to *LHX3* mutation)

| | |
|---|---|
| **MIM** | 600577 |

**Clinical features**  The four patients described so far with *LHX3* mutations have had the following indications: severe growth retardation; complete deficiency of growth hormone, thyrotropin, prolactin, luteinizing hormone, and follicle-stimulating hormone, but normal adrenocorticotropic hormone (ACTH) secretion; and elevated and anteverted shoulders, resulting in a clinical appearance of a stubby neck and severely restricted rotation of the cervical spine. The anterior pituitary was hypoplastic in two patients (see **Figure 1**) and enlarged in one.

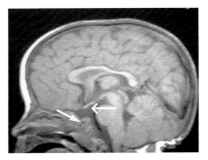

**Figure 1.** Ectopic posterior pituitary (open arrow) and small hypoplastic anterior pituitary gland (closed arrow). Note the absence of the pituitary stalk connecting the pituitary to the hypothalamus.

**Primary tissue or gland affected**  Pituitary.

**Other organs, tissues, or glands affected**  Thyroid, adrenals, gonads (secondary dysfunction due to pituitary deficiency).

**Age of onset**  Infancy or early childhood.

| | |
|---|---|
| **Epidemiology** | This is a rare condition. Only four patients – three siblings from one family and one patient from an unrelated family – have been described. |
| **Inheritance** | Autosomal recessive. |
| **Chromosomal location** | 9q34.3 |
| **Genes** | *LHX3* (LIM homeobox gene 3). |
| **Mutational spectrum** | Two mutations have been described in four patients: one is a missense mutation, and the other a 23-base pair deletion. |
| **Effect of mutation** | LHX3 appears to be involved in the establishment and maintenance of differentiated pituitary cell types. The Y116C missense mutation affects a highly conserved zinc-binding region that functions as a protein-interaction module. The mutation characterized by a 23-base pair deletion, if translated, would result in a severely truncated protein lacking the entire homeodomain. |
| **Diagnosis** | An *LHX3* mutation should be considered in patients with growth failure associated with multiple pituitary hormone deficiencies, preserved ACTH–adrenal function, and restriction of cervical spine rotation. |
| **Notes** | The mechanism responsible for the limitation of neck rotation is not clear. Cervical X-rays failed to reveal malformations of the vertebrae. No abnormalities of muscle or other soft-tissue structures were noted on magnetic resonance imaging. Electromyograms did not reveal evidence of denervation or myopathic changes. |
| **References** | Netchine I, Sobrier ML, Krude H et al. Mutations in LHX3 result in a new syndrome revealed by combined pituitary hormone deficiency. *Nat Genet* 2000;25:182–6. |

# Septo-optical Dysplasia

(also known as: SOD; de Morsier syndrome)

| | |
|---|---|
| **MIM** | 182230 |
| **Clinical features** | Two siblings with a homozygous mutation of the *HESX1* gene have been found. These patients show panhypopituitarism, absence of the septum pellucidum, optic nerve hypoplasia, and agenesis of the corpus callosum. Patients with a heterozygotic mutation show mild pituitary hypoplasia or SOD. The extent of pituitary hormone deficiency may vary from isolated growth hormone (GH) deficiency to panhypopituitarism. Central diabetes insipidus may occur in some patients with SOD. |
| **Primary tissue or gland affected** | Hypothalamus. |
| **Other organs, tissues, or glands affected** | Pituitary; thyroid, adrenals, gonads (secondary dysfunction due to pituitary deficiency) |
| **Age of onset** | Infancy or early childhood. |
| **Epidemiology** | SOD is more prevalent among the offspring of young mothers and in first-born children. Screening of 35 patients with SOD revealed only one set of siblings with a *HESX1* mutation. In a larger series of 228 patients with congenital hypopituitarism of variable severity, three were found to be heterozygotic for *HESX1* mutations. |
| **Inheritance** | Autosomal recessive, autosomal dominant, other. |
| **Chromosomal location** | 3p21.2-p21.1 |
| **Genes** | *HESX1* (homeobox gene expressed in embryonic stem cells; also known as *RPX* [Rathke's pouch homeobox gene]). |

| | |
|---|---|
| **Mutational spectrum** | Two siblings with SOD were found to be homozygous for an Arg53Cys missense mutation. Three different heterozygous missense mutations were identified in patients with milder forms of pituitary hypoplasia or SOD. |
| **Effect of mutation** | Studies in mice indicate that *Hesx1* expression precedes that of *Prop1*. *HESX1* appears to be responsible for the development of Rathke's pouch from the oral ectoderm. The homozygous mutation is unable to bind target DNA. Heterozygotic mutations have reduced affinity for *HESX1* binding elements. |
| **Diagnosis** | SOD should be considered in any child with growth retardation, nystagmus, or impaired vision, and a small optic disc on funduscopic examination. Demonstration of GH deficiency or multiple pituitary hormone deficiencies along with characteristic findings (absence of septum pellucidum or agenesis of the corpus callosum) on magnetic resonance imaging (MRI) confirms the diagnosis. In some patients, an MRI scan may show a hypoplastic or interrupted pituitary stalk and ectopic posterior pituitary. |
| **Counseling issues** | Parents should be aware of the visual impairment in patients with SOD. |
| **Notes** | Agenesis of the corpus callosum has also been found in some patients with a solitary median maxillary central incisor (MIM 147250), a rare dental anomaly associated with variable degrees of holoprosencephaly, hypotelorism, and choanal atresia, along with GH deficiency and/or multiple pituitary hormone deficiencies. Mutations of the sonic hedgehog (*SHH*) gene located at 7q36 have been reported in these patients. |
| **References** | Dattani MT, Martinez-Barbera JP, Thomas PQ et al. Mutations in the homeobox gene HESX1/Hesx1 associated with septo-optic dysplasia in human and mouse. *Nat Genet* 1998;19:125–33. |

Nanni L, Ming JE, Du Y et al. *SHH* mutation is associated with solitary median maxillary central incisor: A study of 13 patients and review of the literature. *Am J Med Genet* 2001;102:1–10.

Parks JS, Brown MR, Hurley DL et al. Heritable disorders of pituitary development. *J Clin Endocrinol Metab* 1999;84:4362–70.

# Isolated Growth Hormone Deficiency

(also known as: IGHD types 1A, 1B, 2, and 3; pituitary dwarfism; GH-releasing hormone deficiency)

| | |
|---|---|
| MIM | 262400 (types 1A and 1B), 173100 (type 2), 307200 (type 3). |
| Clinical features | Signs and symptoms of growth hormone (GH) deficiency from any cause include neonatal hypoglycemia, micropenis, frontal bossing, 'cherubic facies' (see **Figure 2**), short stature with normal body proportions, decreased linear growth rate, and delayed skeletal development (bone age). When exogenous human GH is administered to IGHD type 1A patients, the development of anti-GH antibodies may result in a poor response to this treatment as compared with other types of GH deficiency. IGHD type 1B patients do not usually develop antibodies against exogenous GH. In IGHD with X-linked agammaglobulinemia (type 3), short stature is accompanied by recurrent sinus and pulmonary infections. Neurologic development is normal in most children with IGHD, but prolonged or recurrent hypoglycemia in infancy may result in permanent central nervous system damage. A rare syndrome of familial X-linked mental retardation with IGHD has been described. |
| Primary tissue or gland affected | Anterior pituitary. |
| Age of onset | Birth or early childhood. |
| Epidemiology | The incidence of IGHD is estimated to be 1:4000–10 000 live births. Of patients with IGHD, 5%–30% have affected first-degree relatives, suggesting a possible genetic defect. Since anatomic defects are identified in only 12% of patients with IGHD on magnetic resonance imaging (MRI), mutations not associated with gross anatomic defects may account for GH deficiency in a greater proportion of cases than was previously considered. IGHD type 1B is more frequent than type 1A. Types 2 and 3 are less common. |

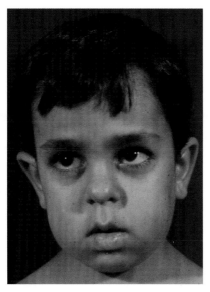

**Figure 2.** A 7-year-old boy with the typical facial features of GH deficiency.

| | |
|---|---|
| **Inheritance** | Autosomal recessive (types 1A and 1B), autosomal dominant (type 2), X-linked (type 3). |
| **Chromosomal location** | 17q22-q24 (types 1A and 2), 7p14-p15 (type 1B), Xq21-q22 (IGHD with X-linked agammaglobulinemia), Xq22-Xq27 (IGHD with X-linked mental retardation). |
| **Genes** | *GH1* (GH types 1A and 2), *GHRH-R* (GH-releasing hormone receptor: type 1B). |
| **Mutational spectrum** | *GH1* gene mutations include deletions, frame shifts, splice-site defects, and nonsense mutations. For the *GHRH-R* gene, missense, nonsense, and splicing mutations have been reported. |
| **Effect of mutation** | In IGHD type 1A, *GH1* mutation results in a total absence of GH synthesis, and undetectable endogenous GH levels in homozygotic patients. In type 1B, the pituitary somatotroph response to GHRH is partially blocked due to either weak binding of GHRH to the |

receptor, abnormal coupling with G-proteins, or interference with protein glycosylation. GH levels in type 1B patients are low, but detectable. Mutations causing IGHD type 2 result in dominant negative expression of the *GH1* gene. Abnormal folding of the mutant GH protein interferes with storage and secretion of wild-type GH produced by the normal allele. In type 2 patients, GH release following standard pharmacological stimuli is impaired, but still present.

**Diagnosis**

Low levels of insulin-like growth factor (IGF)-1, low or undetectable levels of GH in the presence of hypoglycemia (spontaneous or insulin-induced), and lack of response to stimulation by arginine, clonidine, L-dopa, or glucagon confirm the diagnosis of GH deficiency. IGF-1 levels increase following exogenous GH. B-cell number, and immunoglobulins (IgG, IgM, and IgA) may be greatly reduced or absent in patients with X-linked IGHD. Testing for other pituitary hormones is necessary to distinguish isolated GH deficiency from multiple pituitary hormone deficiency. A family history of relatives with IGHD and absence of structural defects on MRI increase the likelihood of a genetic cause.

**Counseling issues**

Early diagnosis and prompt treatment of neonatal hypoglycemia may improve the prognosis for normal neurological development. Early initiation of GH treatment appears to result in a greater final height. Height prognosis is poorer in type 1A patients who develop antibodies to exogenous GH.

**Notes**

Some, but not all, patients with 'Bruton's' X-linked agammaglobulinemia and IGHD have a mutation of the *BTK* gene that encodes a tyrosine kinase. A common molecular mechanism to explain defective B-cell development and GH deficiency remains to be established.

**References**

Binder G, Keller E, Mix M et al. Isolated GH deficiency with dominant inheritance: New mutations, new insights. *J Clin Endocrinol Metab* 2001;86:3877–81.

Procter AM, Phillips JA 3rd, Cooper DN. The molecular genetics of growth hormone deficiency. *Hum Genet* 1998;103:255–72 (Review).

Raynaud M, Ronce N, Ayrault AD et al. X-linked mental retardation with isolated growth hormone deficiency is mapped to Xq22-Xq27.2 in one family. *Am J Med Genet* 1998;76:255–61.

Stewart DM, Notarangelo LD, Kurman CC et al. Molecular genetic analysis of X-linked hypogammaglobulinemia and isolated growth hormone deficiency. *J Immunol* 1995;155:2770–4.

# Growth Hormone Insensitivity

(also known as: Laron-type dwarfism; growth hormone [GH] resistance; primary insulin-like growth factor [IGF]-1 deficiency)

| | |
|---|---|
| MIM | 262500 |
| Clinical features | Patients with GH insensitivity due to GH-receptor defects have normal weight and length at birth, but their postnatal growth velocity is markedly reduced. Bone age is delayed. Craniofacial features include a prominent forehead, hypoplastic nasal bridge, blue sclerae, 'setting sun sign' (where the sclera are visible above the cornea), and prolonged retention of primary dentition. A delay in walking, high-pitched voice, and thin, prematurely aged skin are common. Other characteristics include fasting hypoglycemia, elevated total and low-density lipoprotein cholesterol levels, and a small penis in childhood, but normal genital growth in adolescence. Delayed puberty is common, but reproductive function is normal. In one case, a patient described with IGF-1 deficiency had intrauterine growth retardation, severe postnatal growth failure, sensorineural deafness, and mental retardation. |
| Primary tissue or gland affected | Liver, skeletal epiphyses. |
| Other organs, tissues, or glands affected | All tissues with IGF-1 receptors. |
| Age of onset | Infancy and early childhood. |
| Epidemiology | GH insensitivity is more common in Arabs, oriental or Middle Eastern Jews, and Conversos (Jews converted during the Inquisition). A cohort of 71 patients from Ecuador has been described in which all but one had an E180 splice mutation in the GH-receptor gene. Only one patient with an *IGF-1* gene deletion has been described. |

| | |
|---|---|
| **Inheritance** | Autosomal recessive, autosomal dominant. |
| **Chromosomal location** | 5p13-p12 (GH receptor), 12q22-q24.1 (IGF-1) |
| **Genes** | *GHR* (GH receptor), *IGF-1*. |
| **Mutational spectrum** | Exon deletions, and nonsense, missense, frame-shift, and splice mutations of the *GHR* gene resulting in growth failure have been described. A deletion of two exons was described in the single patient with primary IGF-1 deficiency. |
| **Effect of mutation** | Most of the homozygous and compound heterozygote defects result in absent or very low levels of GH-binding protein (GH-BP), which is the proteolytic product of the extracellular domain of the GH receptor. Three mutations have been described that are associated with normal production and GH-binding of the extracellular domain of the GHR, but failure to dimerize at the cell surface results in a nonfunctional GH receptor. Some mutations interfere with the normal splicing of exon 8, which encodes the transmembrane domain; the resulting truncated GHR protein is able to bind GH, but cannot be anchored to the cell surface. A dominant negative effect was demonstrated for two heterozygous mutations involving exon 9 and intron 9 of the *GHR* gene, resulting in a markedly attenuated intracellular domain. These heterozygotic patients produce both normal and abnormal GHR; the mutant receptor shows increased affinity for GH, but inhibits tyrosine phosphorylation and transcription activation. The single description of a *IGF-1* gene deletion resulted in an IGF-1 peptide truncated from 70 to 25 amino acids. |
| **Diagnosis** | In a patient with severe growth failure, elevated GH levels with markedly reduced IGF-1, IGF-2, and IGF-BP3 serum concentrations, along with elevated levels of IGF-BP1 and IGF-BP2, are diagnostic for GH insensitivity. GH-BP levels are usually low, but normal or elevated GH-BP concentrations may be seen. GH insensitivity should be considered in children with growth retardation, slow |

growth velocity, low (>1 standard deviation below the mean) IGF-1 or IGF-BP3 levels for age, and normal (or exaggerated) GH response to stimulation tests. Hypothyroidism, malnutrition or malabsorption disorders, and chronic systemic diseases must be excluded. Failure of IGF-1 levels to rise after four daily injections of synthetic GH (0.1 mg/kg/day) confirms the diagnosis.

**Counseling issues**

Determination of the mutation responsible for GH insensitivity in a particular patient may be helpful in advising families whether the pattern of inheritance is likely to follow an autosomal recessive or dominant pattern. Treatment with recombinant IGF-1 increases growth velocity in most patients, but hypoglycemia, pain at injection sites, and tachycardia appear to be common side effects.

**Notes**

The dominant negative effect described for heterozygous *GHR* mutations affecting the intracellular domain has not been found in heterozygotes for mutations involving the extracellular or transmembrane domains. The possibility has been raised that mild or partial GH insensitivity due to heterozygosity for *GHR* mutations may be a relatively common cause of idiopathic short stature.

**References**

Rosenbloom AL. Growth hormone insensitivity: Physiologic and genetic basis, phenotype, and treatment. *J Pediatr* 1999;135:280–9 (Review).

Woods KA, Camacho-Hubner C, Savage MO et al. Intrauterine growth retardation and postnatal growth failure associated with deletion of the insulin-like growth factor I gene. *N Engl J Med* 1996;335:1363–7.

# Isolated Central Hypothyroidism

(also known as: isolated thyrotropin deficiency; secondary hypothyroidism; tertiary hypothyroidism; resistance to thyrotropin-releasing hormone [TRH]; TRH deficiency)

| | |
|---|---|
| **MIM** | 275100, 275120 |
| **Clinical features** | Clinical manifestations of isolated central hypothyroidism are similar to those of primary hypothyroidism (see Chapter 3), but are usually mild. A TRH-receptor gene defect presented in an 8-year-old boy with short stature and markedly delayed bone age. |
| **Primary tissue or gland affected** | Pituitary. |
| **Other organs, tissues, or glands affected** | Thyroid. |
| **Age of onset** | Infancy or childhood. |
| **Epidemiology** | Isolated central hypothyroidism of genetic origin is very rare, with an estimated frequency of 0.005% in the general population. Most cases are caused by tumors, infiltrative diseases of the hypothalamus or pituitary, or pituitary atrophy. Rare cases of TSH β-subunit mutations causing central hypothyroidism have been described. Only one patient with resistance to TRH has been reported. |
| **Inheritance** | Autosomal recessive. |
| **Chromosomal location** | 1p13 (TSH β chain), 8q23 (TRH receptor), 3q13.3-q21 (TRH). |
| **Genes** | *TSHB* (TSH β chain), *TRHR* (TRH receptor), *TRH* (TRH gene). |
| **Mutational spectrum** | Mutations reported for the *TSHB* gene are substitutions. The only patient described to date with TRH resistance is a compound |

heterozygote. The maternal allele is a nucleotide substitution, and the paternal allele is a deletion of nine nucleotides. No mutations resulting in TRH deficiency have been described in humans.

**Effect of mutation**
The effects of TSH β mutations in the amino-terminal region include lack of dimer formation, production of a truncated β-subunit, and changes in the amino acid sequence resulting in undetectable circulating TSH levels. Substitutions in the carboxy-terminal region affect biological activity, but do not prevent formation of the heterodimer; therefore, some immunoreactive TSH levels may be detected. In TRH resistance, the maternal allele contains a premature stop codon, resulting in a truncated protein with none of the seven transmembrane domains. The mutated paternal allele presumably alters the tertiary structure of the third transmembrane helix. Mutated receptors show greatly reduced or absent TRH binding *in vitro*.

**Diagnosis**
Serum total and free thyroxine levels are low. Absolute concentrations of TSH may be low, normal, or slightly elevated, but are inappropriately low for the hypothyroxinemia. TSH α-subunit levels may be elevated. The response of TSH to TRH stimulation is typically absent, but a weak, delayed response may be seen. A hyperplastic pituitary was observed on magnetic resonance imaging in one patient with a TSH β mutation, producing high levels of α-glycoprotein subunit.

**Counseling issues**
Screening programs that rely on TSH levels alone will not detect infants with central hypothyroidism.

**Notes**
Because mutations causing isolated TSH deficiency are extremely rare, a careful search for other causes of central hypothyroidism should be performed.

**References**

Bonomi M, Proverbio MC, Weber G et al. Hyperplastic pituitary gland, high serum glycoprotein hormone alpha-subunit, and variable circulating thyrotropin (TSH) levels as hallmark of central hypothyroidism due to mutations of the TSH beta gene. *J Clin Endocrinol Metab* 2001;86:1600–4.

Collu R, Tang J, Castagne J et al. A novel mechanism for isolated central hypothyroidism: Inactivating mutations in the thyrotropin-releasing hormone receptor gene. *J Clin Endocrinol Metab* 1997;82:1561–5.

# Diabetes Insipidus, Central

**(also known as: cDI; neurogenic, neurohypophyseal, pituitary, or cranial DI)**

| | |
|---|---|
| **MIM** | 125700 |
| **Clinical features** | Polydipsia and polyuria with dilute urine are the classical features of cDI. Hypernatremic dehydration occurs when patients with cDI are deprived of water. Presenting symptoms in infancy and early childhood are nonspecific and include poor feeding, failure to thrive, irritability, fever, and constipation. |
| **Primary tissue or gland affected** | Hypothalamic supraoptic and paraventricular nuclei, posterior pituitary. |
| **Other organs, tissues, or glands affected** | Renal tubules. |
| **Age of onset** | Since most cases of cDI are acquired, polyuria and polydipsia may begin at any age. cDI due to congenital cerebral malformations usually appears in infancy. In familial cDI, clinical manifestations usually appear at age 1–6 years. Vasopressin deficiency in familial autosomal dominant cDI may be mild at first, usually progressing to severe cDI over the course of months or years. |
| **Epidemiology** | The incidence of cDI is 3:100 000; approximately 60% are males. Familial cDI accounts for only 1%–2% of all cases of cDI. |
| **Inheritance** | Autosomal dominant, autosomal recessive. |
| **Chromosomal location** | 20p13 |
| **Genes** | *AVP-NPII* (arginine vasopressin-neurophysin II). |

| | |
|---|---|
| **Mutational spectrum** | In-frame deletions, missense, and nonsense mutations have been described for autosomal dominant cDI. A missense mutation was identified in a family with autosomal recessive cDI. |
| **Effect of mutation** | In autosomal dominant cDI, the mutant *AVP-NPII* gene produces a mutant vasopressin precursor, which, due to abnormal folding and lack of self association, is not transported from the endoplasmic reticulum to the Golgi apparatus and secretory vesicles. Mutant preprohormones have abnormal disulfide bridge formation, defective rotational freedom, and/or altered binding of AVP to neurophysin. Gradual accumulation of abnormal vasopressin precursor in the endoplasmic reticulum may interfere with expression of the normal allele by interfering with production of other proteins that are needed to maintain structure and function of the hypothalamic/ neurohypophyseal neurons. The autosomal recessive form of familial cDI is characterized by a mutant form of vasopressin that displays a 30-fold reduction in receptor binding. |
| **Diagnosis** | A 24-hour urine volume >3 L (in adults) distinguishes polyuria from urinary frequency or nocturia. Differential diagnoses include primary polydipsia, compulsive water drinking, gestational DI (due to increased metabolism of vasopressin), diabetes mellitus, chronic renal failure, hypercalcemia, and hypokalemia. Laboratory studies should include blood glucose, creatinine, urea, calcium, and potassium concentrations. Low urine osmolality and elevated serum osmolality in an early morning sample strongly suggest cDI. A water deprivation test, demonstrating failure to produce concentrated urine either in the presence of a 3%–5% weight loss and elevated serum osmolality, or following infusion of hypertonic saline, confirms the diagnosis of cDI. An increase in urine osmolality by >450 mOsm/kg after administration of intranasal 1-desamino-8-D-arginine vasopressin (DDAVP) indicates cDI; a urine osmolality ≤300 mOsm/kg after DDAVP suggests nephrogenic DI. Magnetic resonance imaging (MRI) should be performed to determine whether central nervous system (CNS) tumors or other lesions are present. |

A high-intensity signal or 'bright spot' in the posterior pituitary is seen on T1-weighted MRI in most healthy adults. Absence of the bright spot has been reported in all patients with CNS pathology causing cDI, as well as in most cases of idiopathic cDI. Absence of the T1-weighted bright spot has not been consistently observed in familial cDI.

**Counseling issues**

Nearly all mutations for autosomal dominant familial cDI result in a similar clinical phenotype. Substitutions at NPII57 appear to be associated with an early onset of polyuria; substitutions at the C-terminal cleavage site of the signal protein are associated with late onset of symptoms (age 3–6 years). Since the clinical manifestations of cDI in infants and young children are often nonspecific, the identification of affected children in families with cDI may lead to earlier diagnosis and prompt treatment.

**Notes**

In some patients with idiopathic cDI, MRI scans were initially normal; follow-up MRI over the course of several years subsequently revealed the presence of CNS neoplasia.

**References**

Hansen L, Rittig S, Robertson G. Genetic basis of familial neurohypophyseal diabetes insipidus. *Trends Endocrinol Metab* 1997;8:363–72.

Saborio P, Tipton GA, Chan JC. Diabetes insipidus. *Pediatr Rev* 2000;21:122–9.

Willcutts MD, Felner E, White PC. Autosomal recessive familial neurohypophyseal diabetes insipidus with continued secretion of mutant weakly active vasopressin. *Hum Mol Genet* 1999;8:1303–7.

# Russell–Silver Syndrome

(also known as: RSS)

| | |
|---|---|
| **MIM** | 180860 |
| **Clinical features** | The classical picture of RSS includes intrauterine growth retardation (but normal head circumference), triangular facies, asymmetry (legs are affected more often than arms), and clinodactyly. Feeding problems in infancy and childhood are common, and 'catch-up' growth does not occur. Puberty occurs early and is associated with a poor growth spurt. Boys with RSS often need to undergo surgical repair of hypospadias and inguinal hernia. The average final height is 149.5 cm for boys and 138 cm for girls. |
| **Primary tissue or gland affected** | Bones (short stature, clinodactyly, and dysmorphic facial features). |
| **Age of onset** | Prenatal. |
| **Epidemiology** | The incidence of RSS is unknown, but may be as high as 10%–15% of children with intrauterine growth retardation. |
| **Inheritance** | Sporadic, autosomal dominant. |
| **Chromosomal location** | 17q23-q24,7 |
| **Genes** | Unknown. |
| **Mutational spectrum** | Translocations in the region of 17q25 were noted in several patients with RSS. Uniparental disomy of chromosome 7 has been found in approximately 10% of patients with the typical features of RSS. |
| **Effect of mutation** | Unknown. |

**Diagnosis**            Since maternal uniparental disomy of chromosome 7 is found in only a small minority of children with RSS, diagnosis continues to be based on physical findings and growth pattern.

**Notes**                Growth hormone treatment may be effective in improving the final height of RSS children. At this time, there is no consensus regarding the optimum dosage or duration of growth hormone treatment.

**References**           Price SM, Stanhope R, Garrett C et al. The spectrum of Silver–Russell syndrome: A clinical and molecular genetic study and new diagnostic criteria. *J Med Genet* 1999;36:837–42.

# Noonan's Syndrome

(also known as: NS; male Turner syndrome; Turner phenotype with normal karyotype)

| | |
|---|---|
| MIM | 163950 |
| Clinical features | NS is characterized by short stature, dysmorphic facial features, and congenital cardiovascular anomalies. The typical facial appearance of NS patients consists of a broad forehead, hypertelorism, ptosis, down-slanting palpebral fissures, high-arched palate, and low-set, posteriorly rotated ears. Nearly 90% of NS patients have a cardiac defect, including pulmonic stenosis, hypertropic cardiomyopathy, atrioventricular septal defects, and coarctation of the aorta. Other features include a short and webbed neck, pectus excavatum, pectus carinatum, widely spaced nipples, spine deformities, cubitus valgus, mental retardation, cryptorchidism, and bleeding diathesis. IQ is variable: approximately 15% of NS patients have significant mental retardation. |
| Primary tissue or gland affected | Facial bones and cranium, skeleton, cardiovascular system. |
| Age of onset | Cardiac anomalies may be suspected on prenatal ultrasound examination. Other features are recognized in infancy, childhood, or early puberty. |
| Epidemiology | The estimated prevalence of NS is 1:1000–2500 live births. |
| Inheritance | Sporadic, autosomal dominant, autosomal recessive. |
| Chromosomal location | 12q24.1 |
| Genes | PTPN11 (nonreceptor protein tyrosine phosphatase, SHP2). |
| Mutational spectrum | All of the mutations found in NS are missense exonic changes. Most are clustered in exons 3 and 8. |

**Effect of mutation**

SHP2 is a cytosolic protein tyrosine phosphatase. In animal models, SHP2 is involved in mesodermal patterning, limb development, hematopoietic cell differentiation, and semilunar valvulogenesis. It is also a key molecule in the cellular response to growth factors, hormones, cytokines, and cell adhesion molecules. The *PTPN11* mutations in NS appear to cause gain of function effects by stabilizing SHP2 in the active conformation.

**Diagnosis**

Until the association with the *PTPN11* mutation was discovered, diagnosis of NS depended on the recognition of typical clinical features together with a normal karyotype. The most characteristic features of NS are short stature, webbed neck, pectus deformities, ptosis, and right-sided congenital heart disease (in contrast to the left-sided cardiac anomalies seen in Turner's syndrome). Identification of a *PTPN11* mutation can help to confirm the diagnosis in patients in whom the clinical features are less obvious. Differential diagnoses include fetal alcohol syndrome and prenatal exposure to the anticonvulsants primidone and hydantoin.

**Counseling issues**

In contrast to Turner's syndrome, ovarian function, including fertility, is normal in women with NS. Men with NS have delayed puberty; cryptorchidism is common, and is associated with hypoplastic testes and germinal cell aplasia. Nevertheless, approximately 50% of NS males have normal testicular function and fertility.

**Notes**

Testosterone treatment may be needed at puberty, especially in NS males with a history of cryptorchidism.

**References**

Tartaglia M, Kalidas K, Shaw A et al. PTPN11 mutations in Noonan syndrome: Molecular spectrum, genotype–phenotype correlation, and phenotypic heterogeneity. *Am J Hum Genet* 2002;70:1555–63.

# Turner's Syndrome

(also known as: TS; gonadal dysgenesis; Ullrich–Turner's syndrome)

**MIM**          Not listed.

**Clinical features**          TS patients are unambiguously phenotypic females. In most instances, spontaneous puberty and menstruation do not occur, but 25%–30% of patients may undergo varying degrees of pubertal development, including menses. The final height prognosis appears to be worse in TS girls who undergo spontaneous puberty. Short stature is the most common somatic feature of TS. Growth retardation *in utero*, during infancy, and throughout childhood, along with a decreased or absent pubertal growth spurt, result in a mean final height of 143 cm. Skeletal anomalies include mesomelia, micrognathia, cubitus valgus, short metacarpals, and Madelung deformity (shortened and bowed radius with subluxation of the distal ulna; see **Figure 3**).

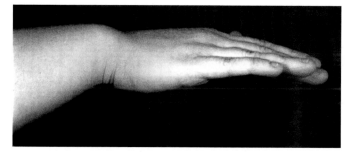

**Figure 3.** Madelung deformity in a woman with Leri–Weill dyschondrosteosis. The same deformity is frequently found in Turner's syndrome.

Other characteristic features include webbing of the neck, low posterior hairline, prominent ears, epicanthal folds, broad ('shield') chest with an appearance of widely-spaced nipples, hyperconvex fingernails, and nail hypoplasia. Lymphedema of the hands and feet are common in infancy and usually resolve spontaneously over months to years. Bicuspid aortic valve and coarctation of the aorta are the most common cardiovascular defects, though hypertension tends to appear even in the absence of structural cardiovascular disease. Hearing loss is due to increased middle ear infections in childhood and sensorineural hearing loss in adults. Hypothyroidism develops in 25%–30% of adult TS women, and the prevalence of type 2 diabetes mellitus is increased more than 4-fold in TS patients. Osteoporosis is common, but less severe in patients who receive growth hormone (GH) treatment in childhood. Overall intelligence is not impaired, but selective defects in visual–spatial information processing, arithmetic, and nonverbal skills have been noted.

| | |
|---|---|
| **Primary tissue or gland affected** | Gonads. |
| **Other organs, tissues, or glands affected** | Lymph vessels, heart and blood vessels, bone, kidneys. |
| **Age of onset** | Somatic manifestations appear *in utero*. Accelerated ovarian senescence leads to ovarian failure in most 45,X patients by late infancy. When somatic anomalies are mild or absent, the diagnosis may be reached during evaluation of short stature in childhood, or amenorrhea and delayed puberty in adolescence or early adulthood. |
| **Epidemiology** | TS affects 1:2000–5000 female live births. The true prevalence of TS *in utero* is much higher (approximately 1%–2%), but nearly 99% of TS embryos and fetuses spontaneously abort in the first or second trimester. |
| **Inheritance** | TS is sporadic, but rare cases of familial TS, associated with an inherited X-chromosome anomaly or hereditary mosaicism, have been described. |
| **Chromosomal location** | X chromosome. |

| Genes | Genes located at sites on both arms of the X chromosome appear to be associated with the various gonadal and somatic features of this disorder. Candidate genes for specific aspects of TS include: *SHOX* (short-stature homeobox) located at Xp22.33; the transcription factor ZFX (zinc finger X) at Xp11.1-p22; lymphogenic genes in the region of Xp11.3; *USP9X* (also known as *DFRX* [drosophila fat facets related X]) located in Xp11.4; and *RPS4X* (ribosomal protein S4) and *DIAPH2* (the homolog of drosophila diaphanous), both found on the long arm of chromosome X. |
|---|---|
| Mutational spectrum | Karyotypes of peripheral blood lymphocytes show complete loss of one X chromosome in approximately 50% of TS women. Other karyotypes include partial absence of one X chromosome or mosaicism. Some mosaic forms of TS include a Y chromosome. The paternal X chromosome is lost during meiosis in 60%–80% of TS women. |
| Effect of mutation | Proposed mechanisms by which the TS phenotype appears include haploinsufficiency of a gene (or genes) expressed in both sex chromosomes, or imprinting that leads to the expression of a gene from only one parent. During human embryogenesis, the *SHOX* gene is expressed primarily in the elbow, knee, and pharyngeal arches. *SHOX* defects may account for TS anomalies such as micrognathia, mesomelia, cubitus valgus, high-arched palate, short metacarpal, and the Madelung deformity. The transcription factor ZFX may also contribute to the short stature of TS individuals. Lymphedema, webbed neck, and congenital heart disease may all be caused by abnormal development of the lymphatic vessels under the control of a proposed lymphogenic gene. *RSP4X* resides in the Xq region associated with lymphedema and poor survival *in utero*. *USP9X* and *DIAPH2* may account for premature ovarian failure. |
| Diagnosis | TS may be suspected or diagnosed on prenatal ultrasound when cystic hygroma of the neck, fetal edema, coarctation of the aorta, renal anomalies, and/or growth retardation are noted. Elevated α-fetoprotein, human chorionic gonadotropin, unconjugated estriol, |

and inhibin A are maternal serum markers associated with an increased risk of fetal abnormalities, including TS. Prenatal karyotypes do not always correlate with postnatal findings; chromosomal studies should be repeated postnatally to confirm the diagnosis of TS, especially when the results of amniocentesis suggest mosaicism. Serum FSH levels are typically elevated in adolescent girls and adult women with TS. Serum gonadotropin levels in infancy and childhood are not helpful; gonadotropin levels are elevated in normal infants, and are relatively low during childhood, even in girls with classical TS. Buccal smears have no role in screening for suspected TS; the confirmation or exclusion of TS requires determination of the karyotype.

**Counseling issues**

Caution is necessary in interpreting the results of a fetal karyotype for the purpose of prenatal counseling. Even for a nonmosaic 45,X karyotype, there is a very wide range of expression for the typical somatic features of TS. Most cases of 45,X/46,XY mosaicism detected in prenatal screening result in phenotypically normal male infants. The risk of TS is not increased with advanced maternal age, and there is no increased risk of recurrence for subsequent pregnancies. The risk of gonadoblastoma is estimated to be 10%–30% in mosaic TS patients with a Y chromosome or Y-chromosomal fragment. Probing for Y-chromosomal material is not routinely indicated in all TS patients, but should be performed if virilization is present, or if a sex-chromosomal fragment of unknown origin is observed. Probing for the *SRY* gene may be helpful in identifying a Y-chromosomal fragment; however, *SRY* does not appear to be responsible for the development of gonadoblastoma. Infertility due to ovarian failure is a hallmark of TS, but rare cases of spontaneous pregnancies and healthy offspring have been described, including in some women with a 45,X karyotype. *In vitro* fertilization has shown some success using ova or embryo donors, as well as cryopreservation of the patient's own oocytes when residual ovarian function is present. Patients with TS are at increased risk for dissection of the aorta – a high index of suspicion for this potentially fatal complication is essential in TS women presenting to primary care physicians or emergency rooms with complaints of chest or abdominal pain. When all karyotypes causing TS are included, the mean life expectancy is 69 years; for women with a 45,X karyotype, the mean life expectancy is 64.4 years.

**Notes**

Since short stature may be the major, or only, presenting feature of TS, chromosomal studies should be part of the routine evaluation of short stature in girls to rule out TS. GH alone or together with oxandrolone increases the final height by 5–7 cm or more. Early initiation of GH treatment permits a longer duration of therapy before estrogen replacement (and consequent epiphyseal closure) is begun.

**References**

Health supervision for children with Turner syndrome. American Academy of Pediatrics. Committee on Genetics. *Pediatrics* 1995;96:1166–73.

Ranke MB, Saenger P. Turner's syndrome. *Lancet* 2001;358:309–14 (Review).

# Short Stature Associated with *SHOX* Mutations

(also known as: idiopathic short stature [ISS]; Leri–Weill dyschondrosteosis [LWD]; Langer mesomelic dysplasia [LMD])

| | |
|---|---|
| **MIM** | 604271 (ISS), 127300 (LWD), 249700 (LMD) |

**Clinical features**

LWD, a form of mesomelic dysplasia, is characterized by short stature, forearms, and lower legs, and Madelung deformity (shortened and bowed radius with subluxation of the distal ulna). LMD is a rarer and clinically more severe form of growth retardation. In LMD, distal hypoplasia of the ulna and proximal hypoplasia or aplasia of the femur are manifestations of a homozygous *SHOX* defect. In children with ISS, short stature without obvious features of skeletal dysplasia may be the only clinical sign of a *SHOX* mutation. Turner's syndrome (TS) is caused by loss of one copy of the X chromosome, and haploinsufficiency of the *SHOX* gene is thought to cause some of the skeletal abnormalities found in TS. For a more detailed presentation of TS, the reader is referred to the previous entry in this chapter.

**Primary tissue or gland affected**

*SHOX* gene expression in humans is greatest in embryonic osteogenic tissue, and in the first and second pharyngeal pouches.

**Age of onset**

*SHOX* mutations affect the developing skeleton *in utero*.

**Epidemiology**

The female to male ratio among patients with LWD is approximately 3:1. Clinical manifestations of LWD tend to be more severe in women. Because LWD is milder in males, it is possible that this condition is underdiagnosed in boys, and that some males thought to have ISS may actually have mild cases of LWD. The prevalence of *SHOX* mutations among patients with LWD is 60%–100%. *SHOX* mutations were found in approximately 2% of children with ISS (defined as height below the third percentile), implying that the overall prevalence of a *SHOX* gene defect in the general population may be greater than 1 in 2000. If this estimate is accurate,

SHOX disorders could be as common as growth hormone (GH) deficiency or TS.

**Inheritance**    Autosomal dominant.

**Chromosomal location**    Xpter-p22.32, Ypter-p11.2

**Genes**    *SHOX* (short-stature homeobox gene).

**Mutational spectrum**    Deletions, frame shifts, and nonsense mutations are found in LWD and in some patients with ISS. Homozygous deletions or point mutations result in LMD. Haploinsufficiency of *SHOX* due to the loss of one X chromosome is found in TS.

**Effect of mutation**    *SHOX* is located on the pseudoautosomal region (PAR1) of the sex chromosomes, and, along with other genes located in this region, does not undergo X inactivation. Pairing occurs during meiosis, and thus recombination between X and Y chromosomes is possible. Two functional copies are normally present in both males and females. *SHOX* is a 40-kb gene comprised of seven exons that encode two transcripts resulting from alternative splicing mechanisms. The encoded protein appears to be a transcription factor essential for normal growth and skeletal development. Both deletions and point mutations are associated with similar phenotypes in LWD; the point mutations most likely result in loss of function or exert a dominant negative effect. Loss of one X chromosome in TS results in haploinsufficiency of *SHOX*. Homozygous deletions or point mutations result in LMD.

**Diagnosis**    DNA testing for *SHOX* mutations is indicated in patients with characteristic physical findings associated with *SHOX* defects such as Madelung deformity, mesomelic dysplasia, or other skeletal features usually seen in TS. When normal laboratory screening tests, karyotype determination in girls, and GH-stimulation tests result in a diagnosis of 'idiopathic short stature', *SHOX* testing should be considered, even in the absence of typical skeletal deformities.

| | |
|---|---|
| **Counseling issues** | The identification of *SHOX* defects among children with ISS might eventually have important therapeutic implications, if it can be shown that response to GH treatment in ISS due to *SHOX* haploinsufficiency is similar to that observed in girls with TS. |
| **Notes** | An 'overdose' of *SHOX* genes might explain some of the features of multiple X-chromosome disorders, such as tall stature and long limbs in Klinefelter's syndrome. |

**References**

Blaschke RJ, Rappold GA. SHOX: Growth, Leri–Weill and Turner syndromes. *Trends Endocrinol Metab* 2000;11:227–30 (Review).

Rappold GA, Fukami M, Niesler B et al. Deletions of the homeobox gene SHOX (short stature homeobox) are an important cause of growth failure in children with short stature. *J Clin Endocrinol Metab* 2002;87:1402–6.

Ross JL, Scott C Jr, Marttila P et al. Phenotypes associated with SHOX deficiency. *J Clin Endocrinol Metab* 2001;86:5674–80.

# Bardet–Biedl Syndrome

(also known as: BBS)

| | |
|---|---|
| **MIM** | 209900 |
| **Clinical features** | Major (primary) clinical features of BBS include retinal dystrophy of the photoreceptors (rod–cone dystrophy), polydactyly, obesity, learning disorders, hypogonadism, and renal anomalies. Minor (secondary) features include speech delays or disturbances, strabismus, brachydactyly, syndactyly, developmental delays, polyuria and polydipsia due to nephrogenic diabetes insipidus, ataxia and poor muscle co-ordination, diabetes mellitus, high-arched palate and crowding of teeth, congenital heart disease, and hepatic fibrosis. Hypothalamic–pituitary dysfunction and/or primary gonadal failure have been implicated in causing the hypogonadism. Defects of the female genital tract may include hypoplasia of the Fallopian tubes, uterus, and ovaries, and vaginal atresia. |
| **Primary tissue or gland affected** | Eyes, hands, feet, adipose tissue, testes, kidneys. |
| **Other organs, tissues, or glands affected** | Teeth, heart, liver. |
| **Age of onset** | Limb deformities are present *in utero*. Excessive weight gain begins in early childhood, and is usually noted by 1 year of age. Retinal dystrophy may appear in childhood, but is found in nearly all patients by the age of 20 years. |
| **Epidemiology** | In North America and Europe the incidence is 1:140 000–160 000. Presumably due to a founder effect, BBS is more common in Kuwait and Newfoundland, with rates of 1:13 500 and 1:17 500, respectively. |

| | |
|---|---|
| **Inheritance** | Studies of the familial occurrence of BBS indicate an autosomal recessive pattern. However, analysis of phenotype–genotype associations suggests that BBS is not a single-gene recessive disease, but rather a complex syndrome that requires at least three mutant alleles of two or more genes to produce the BBS phenotype. Terminology used to describe this hereditary pattern includes 'digenic' and 'triallelic' inheritance, and 'recessive inheritance with a modifier of penetrance'. |
| **Chromosomal location** | 11q13 (BBS1 locus), 16q21 (BBS2), 3p13 (BBS3), 15q22.3-q23 (BBS4), 2q31 (BBS5), 20p12 (BBS6). |
| **Genes** | Six BBS loci have been identified by linkage association. An additional locus (BBS7), whose chromosomal location is unknown, has been proposed to explain the occurrence of the BBS phenotype in pedigrees without identifiable mutations in known BBS loci. Mutations in *BBS1* account for 40%–50% of all cases of the syndrome. The relative contributions of *BBS2*, *BBS3*, *BBS4*, *BBS5*, and *BBS6* are 8%–16%, 2%–4%, 1%–3%, 3%, and 4%–5%, respectively. The 20%–40% of cases without an identifiable mutation in any of the first six *BBS* genes have been assigned to the as yet uncharacterized *BBS7* gene. The first *BBS* gene to be characterized was *BBS6*, which coincides with the *MKKS* (McKusick–Kaufman syndrome) gene. The *BBS6/MKKS* gene encodes a type 2 class of chaperonins that have a role in ATP-dependent facilitation of protein folding. Structural analysis of the BBS2 gene product predicts an N-terminal coiled domain; the functional role of this protein is unknown. Similarly, the function of the recently cloned BBS1 protein is not known. The BBS4 protein closely resembles O-linked *N*-acetylglucosamine transferase, and is thought to modulate cellular responses to extracellular stimuli. |
| **Mutational spectrum** | Frame shifts, missense, or splice mutations. |

| **Effect of mutation** | Homozygous or compound heterozygous frame-shift mutations result in premature stop codons that produce either truncated protein or nonsense RNA, resulting in no protein production. Loss of function may be milder for some *BBS2* missense mutations. Total loss of BBS6 function is associated with clinical BBS, whereas alleles causing milder degrees of impairment may be associated with MKKS. |
|---|---|
| **Diagnosis** | The proposed clinical criteria for diagnosing BBS require either four 'primary' features, or three 'primary' and two 'secondary' features, as described above. |
| **Counseling issues** | Fertility is very rare in men; only two cases of BBS men fathering children have been reported. Despite the genital anomalies often seen in BBS women, the chances of achieving fertility are somewhat better than for men. |
| **Notes** | The term 'Laurence–Moon–Bardet–Biedl' has, in some reports, been incorrectly applied to patients with BBS. The patients described by Laurence and Moon appear to represent a different syndrome, characterized by paraplegia, but without polydactyly or obesity. Other conditions in which mutations at more than one locus are necessary to produce the disease phenotype include some types of retinitis pigmentosa and dominant Hirschsprung's disease. |
| **References** | Katsanis N, Lupski JR, Beales PL. Exploring the molecular basis of Bardet–Biedl syndrome. *Hum Mol Genet* 2001;10:2293–9 (Review). |
| | Katsanis N, Ansley SJ, Badano JL et al. Triallelic inheritance in Bardet–Biedl syndrome, a Mendelian recessive disorder. *Science* 2001;293:2256–9. |
| | Mykytyn K, Nishimura DY, Searby CC et al. Identification of the gene (BBS1) most commonly involved in Bardet– Biedl syndrome, a complex human obesity syndrome. *Nat Genet* 2002;31:435–8. |

# Prader–Willi Syndrome

(also known as: PWS)

| | |
|---|---|
| MIM | 176270 |
| Clinical features | PWS is characterized by short stature, hypotonia, excessive appetite with progressive obesity, hypogonadism, mental retardation, behavior abnormalities, sleep disorders (including apnea), and dysmorphic features. Feeding problems in infancy may result in failure to thrive, with increased appetite and weight gain beginning at age 1–6 years. Hypogonadotropic hypogonadism may result in cryptorchidism, small penile size, and delayed puberty. Behavioral problems include aggressiveness, obsessive–compulsive disorders, itching and skin picking, and relative insensitivity to pain. Among the dysmorphic features are small hands and feet, almond-shaped eyes, strabismus, and a narrow bifrontal diameter. Scoliosis and kyphosis are common. |
| Primary tissue or gland affected | Central nervous system (CNS), pituitary, bones. |
| Other organs, tissues, or glands affected | Gonads. |
| Age of onset | Prenatal ultrasound may detect decreased fetal movement. More often, PWS is diagnosed in infancy or early childhood, though some patients may be recognized in adolescence or adulthood. |
| Epidemiology | Prevalence is estimated at 1:10 000–16 000. |
| Inheritance | The risk of recurrence for typical deletions or maternal disomy (approximately 98% of PWS) is negligible. For parental translocations (<2% of PWS cases), the recurrence risk is significant and prenatal testing is possible. In the rare cases of an imprinting center mutation, the recurrence risk is about 50%; prenatal testing is possible only if the proband defect is known. |

| | |
|---|---|
| **Chromosomal location** | 15q11-q13 |
| **Genes** | *SNRPN* (small ribonucleoprotein N), *NDN* (necdin; this protein is expressed in the nuclei of neurons, mainly in the CNS). |
| **Mutational spectrum** | Lack of expression of imprinted paternal genes is the common genetic defect leading to PWS. About 70% of patients have a paternal deletion in the 15q11-q13 region. Maternal uniparental disomy is found in approximately 25% of affected patients. In 2%–5% of patients, biparental inheritance of chromosome 15 is observed; abnormal methylation affects gene expression and imprinting in these individuals. Translocations are noted in about 1% of PWS cases. Deletions in the maternal 15q11-q13 region are associated with Angelman syndrome (MIM 105830), which has a very different phenotype to that of PWS. Typical features of Angelman syndrome include severe mental retardation, absence of intelligible speech, ataxic gait, paroxysms of laughter, and seizure disorders; in contrast to PWS, obesity is not an integral feature of Angelman syndrome. |
| **Effect of mutation** | SNRPN protein (SmN) is involved in RNA splicing. Necdin is a nuclear protein involved in cerebral development. Expression of *NDN* appears to be limited to the paternal allele. |
| **Diagnosis** | Parent of origin-dependent methylation analysis will diagnose almost all PWS patients, regardless of the etiology of the condition. To further pursue the etiology of the syndrome, segregation analysis of polymorphic short tandem repeat sequences from 15q11-q13 in the patient and parents can differentiate between cases of uniparental disomy (UPD) or deletion. Fluorescence *in situ* hybridization (FISH) analysis, using the commercially available DNA probes from the PWS region, can identify the common deletion; however, it should be noted that some patients have smaller deletions that can be missed in the FISH analysis. In patients with a deletion, it is recommended that cytogenetic analysis should be performed in case |

the deletion is secondary to an unbalanced translocation (inherited or *de novo*). With the widespread availability of genetic testing to confirm PWS, clinical diagnostic criteria are important in raising clinical suspicion in order to ensure that all appropriate subjects are referred for genetic testing. Because the clinical features of PWS evolve with age, new diagnostic criteria, based on the age at assessment, have been proposed. These new criteria emphasize: hypotonia with poor suckling in infancy; developmental delay at age 2–6 years; hyperphagia and obesity at age 6–12 years; and mental retardation, behavior problems, obesity, and hypogonadism from age 13 years through to adulthood. Differential diagnoses include Cushing's syndrome, Bardet–Biedl syndrome (MIM 209900, previous entry), and Alstrom's syndrome (MIM 203800, p. 193). The clinical features of Bardet–Biedl syndrome include obesity, mental retardation, retinitis pigmentosa, polydactyly, and hypogonadism. This syndrome follows a recessive pattern of inheritance, is linked to six distinct loci, and requires mutations in two separate loci (digenic recessive). Alstrom's syndrome is indicated by obesity, retinitis pigmentosa, and deafness, but is not associated with mental retardation, polydactyly, or hypogonadism.

**Counseling issues**

Early diagnosis permits families to begin intensive dietary and behavior management before marked obesity develops. Physical therapy may improve muscle tone and help to prevent scoliosis. The risk of recurrence is low (approximately 1%) in families where the proband has a deletion or UPD. Translocations are associated with a higher rate of recurrence. Prenatal molecular diagnosis can be performed on chorionic villus samples.

**Notes**

Growth hormone treatment of children with PWS improves longitudinal growth, and, in conjunction with a strict diet, reduces fat mass and increases muscle mass. Improvements in motor performance, endurance, and physical appearance may also improve psychosocial functioning. Although adverse effects do not appear to be greater than those observed during growth hormone treatment of non-PWS children, close monitoring of glucose tolerance and scoliosis is especially indicated. Sex hormone replacement is controversial. Cryptorchidism should be corrected in order to prevent or detect testicular malignancies. Low-dose testosterone substitution in adolescent boys and young adult men may improve strength and endurance, but can also exacerbate aggressive behavior. Estrogen replacement in women may be beneficial in preventing osteoporosis and improving quality of life, though endogenous peripheral conversion of adrenal androgens to estrogens in obese women may achieve the same goals.

**References**

Burman P, Ritzen EM, Lindgren AC. Endocrine dysfunction in Prader–Willi syndrome: A review with special reference to GH. *Endocr Rev* 2001;22:787–99 (Review).

Fridman C, Varela MC, Kok F et al. Prader–Willi syndrome: Genetic tests and clinical findings. *Genet Test* 2000;4:387–92.

Gunay-Aygun M, Schwartz S, Heeger, S et al. The changing purpose of Prader–Willi syndrome clinical diagnostic criteria and proposed revised criteria. *Pediatrics* 2001;108:E92.

# Pallister–Hall Syndrome

(also known as: PHS; hypothalamic hamartoma; hypopituitarism; polydactyly; imperforate anus)

| | |
|---|---|
| **MIM** | 146510 |
| **Clinical features** | Hypothalamic hamartoma disrupts pituitary development and leads to a wide range of pituitary dysfunction. Some patients are asymptomatic, while others have multiple pituitary hormone deficiencies. Precocious puberty occurs in some cases. Craniofacial anomalies include a flat nasal bridge, low-set ears, cleft palate, cleft uvula, and bifid epiglottis. Polydactyly, syndactyly, and nail dysplasia of the fingers and toes may be seen. Micropenis and cryptorchidism are consequences of hypopituitarism. Renal anomalies (agenesis, hypoplasia, or ectopia), congenital heart defects (patent ductus, ventricular septal defect, endocardial cushion defects, mitral and aortic valve anomalies, and aortic coarctation), and imperforate anus have been described. |
| **Primary tissue or gland affected** | Hypothalamus. |
| **Other organs, tissues, or glands affected** | Pituitary, epiglottis, kidneys, anus, distal extremities, nails. |
| **Age of onset** | The onset of PHS is prenatal. The diagnosis is often reached in infancy, but may present later in childhood or in adolescence. Asymptomatic hypothalamic hamartomas are sometimes detected in the parents of an affected child. |
| **Epidemiology** | PHS is rare, but the syndrome is being recognized with increasing frequency since the original report by Hall in 1980. |
| **Inheritance** | Autosomal dominant. |

| | |
|---|---|
| **Chromosomal location** | 7p13 |
| **Genes** | *GLI3* (glioma-associated oncogene homology 3). |
| **Mutational spectrum** | PHS is caused by frame-shift mutations. |
| **Effect of mutation** | GLI3 is a zinc finger transcription factor. Mutations in *GLI3* produce a truncated protein that lacks a critical functional domain. |
| **Diagnosis** | The magnetic resonance imaging (MRI) appearance of the hypothalamic hamartoma (a noncalcified, nonenhancing, homogenous mass) usually differentiates this lesion from more common suprasellar lesions such as craniopharyngioma and glioma. The presence of a bifid epiglottis is strongly linked to PHS. A thorough evaluation of pituitary function and search for congenital defects in other organ systems must be performed. Visual field testing and serial MRI scans are indicated. |
| **Counseling issues** | Parents of affected patients should be evaluated for possible asymptomatic hypothalamic hamartomas. Recognition of autosomal dominant inheritance in familial cases should permit early diagnosis and treatment of the endocrinologic, cardiac, and urogenital complications of this syndrome. |
| **Notes** | PHS is an important reminder that not all central nervous system space-occupying lesions in patients with pituitary dysfunction require neurosurgical intervention. Microdeletions of the region of chromosome 7 that include the *GLI3* gene are associated with a different dominant syndrome, called Greig cephalopolysyndactyly (MIM 175700). |
| **References** | Biesecker LG, Abbott M, Allen J et al. Report from the workshop on Pallister–Hall syndrome and related phenotypes. *Am J Med Genet* 1996;65:76–81. |

Hall J, Pallister PD, Clarren SK et al. Congenital hypothalamic hamartoblastoma, hypopituitarism, imperforate anus, and postaxial polydactyly – a new syndrome? Part I: clinical, causal, and pathogenetic considerations. *Am J Med Genet* 1980;7:47–74.

Kuo JS, Casey SO, Thompson L et al. Pallister–Hall syndrome: Clinical and MR features. *Am J Neuroradiol* 1999;20:1839–41.

# Marfan's Syndrome

(also known as: MFS)

| | |
|---|---|
| **MIM** | 154700 |

| | |
|---|---|
| **Clinical features** | Tall stature and increased arm span are characteristic features. Children and adolescents with MFS tend to be underweight for their height. Other skeletal findings include pectus excavatum or carinatum, joint laxity and hypermobility, arachnodactyly, scoliosis, thoracic lordosis, crowded teeth associated with a high-arched palate and narrow jaw, and flat feet. Ocular manifestations include ectopia lentis (lens subluxation), myopia, increased axial globe length, and corneal flatness. The most common cardiovascular findings are mitral valve prolapse, aortic root dilatation, mitral and aortic regurgitation, and dissection and rupture of the aorta. Widening of the dural sac ('dural ectasia'), usually at the lower end of the lumbosacral spinal column, presents as back pain and headaches. Dural ectasia can be found in >60% of MFS patients and is considered to be a major criterion for the diagnosis. Other characteristic features include spontaneous pneumothorax, recurrent hernia, and striae atrophicae ('stretch marks'). When MFS presents in infancy, the clinical manifestations are usually more severe than in older children. |
| **Primary tissue or gland affected** | Eyes, bones, heart, blood vessels. |
| **Age of onset** | MFS may appear as a neonatal or infantile form, or, more commonly, later in childhood or during early adolescence. |
| **Epidemiology** | The prevalence of MFS is estimated to be 2–3:10 000. |
| **Inheritance** | Autosomal dominant. At least 25% of patients have parents who do not have MFS; these cases represent new mutations. |

| Chromosomal location | 15q21.1 |
|---|---|

**Genes**               *FBN1* (fibrillin 1).

**Mutational spectrum**    Several hundred mutations of *FBN1*, mainly missense mutations and frame shifts, have been described.

**Effect of mutation**    FBN1, a 350-kD glycoprotein, is the major component of microfibrils, which are present in connective tissue (including bone). The *FBN1* gene is 110-kb long and contains 65 exons. Missense mutations affect cysteine residues in the epidermal growth factor (EGF)-like domains of FBN1 and/or interfere with calcium binding in the EGF domain. Mutant FBN monomers combine with normal monomers to form defective FBN multimers, characteristic of a dominant negative effect. Missense mutations in exons 24–26 or deletions of exons 31 or 32 are associated with the severe, infantile form of MFS. Phenotypic manifestations of frame-shift mutations are variable; the severity of the phenotype appears to correlate with the amount of mutant FBN1 expressed. The same mutation can show a very wide range of phenotypic expression.

**Diagnosis**    When no other family members are affected, major criteria in at least two organ systems and minor involvement of a third organ system are required for diagnosis. The most consistent findings are tall stature and evidence of aortic root dilatation on echocardiography. A slit-lamp examination should be performed to look for ocular abnormalities. Urine amino acid screening should exclude homocystinuria. The diagnostic criteria are less stringent in cases with a positive family history. Other conditions that bear resemblance to some of the features of MFS are Ehlers–Danlos syndrome, multiple endocrine neoplasia type 2B (MIM 162300, p. 9), and Klinefelter's syndrome (p. 330). Acromegaly and gigantism, Sotos' syndrome, Weaver syndrome, and precocious puberty in young children need to be considered as differential diagnoses of tall stature.

**Counseling issues**    Prenatal ultrasound is not helpful in the first two trimesters, but chorionic villus biopsy and amniocentesis can lead to prenatal diagnosis when a specific mutation has been identified in the affected parent, or when linkage around markers near the *FBN1* locus can be established. Preimplantation diagnosis has been reported in some cases. Sporadic cases of MFS are associated with older paternal age (36 years vs. 29 years). Early diagnosis, appropriate follow-up examinations, and timely intervention to prevent life-threatening cardiovascular complications are required to increase life expectancy and improve quality of life for these patients.

**Notes**    Although MFS is not, in a strict sense, a hormonal disorder, patients are commonly referred to an endocrinologist for evaluation of tall stature and suspected growth disorder. Clinicians should be familiar with this condition, and be able to diagnose MFS and refer patients for appropriate treatment of the cardiac, ocular, and skeletal manifestations.

**References**    Giampietro PF, Raggio C, Davis JG. Marfan syndrome: Orthopedic and genetic review. *Curr Opin Pediatr* 2002;14:35–41 (Review).

Pyeritz RE. The Marfan syndrome. *Annu Rev Med* 2000;51: 481–510 (Review).

# Obesity due to Leptin Deficiency or Resistance

(also known as: leptin deficiency; leptin receptor defect)

| | |
|---|---|
| **MIM** | 164160 (leptin), 601007 (leptin receptor) |
| **Clinical features** | Children with congenital leptin deficiency are a normal weight at birth, but rapidly gain weight. Obesity is noted by age 3–4 months, becomes increasingly severe with age, and is accompanied by hyperphagia and hyperinsulinemia. Hypogonadotropic hypogonadism is present in adult patients. Patients with a leptin receptor mutation have elevated serum leptin levels. These patients have normal birth weights, but become severely obese within the first few months of life, and display abnormal eating behavior, impulsiveness, and social disability that resembles some of the psychological abnormalities seen in Prader–Willi syndrome, but without mental retardation. In patients with the receptor defect, growth retardation, lack of pubertal development, and central hypothyroidism associated with low levels of growth hormone secretion, gonadotropins, thyrotropin, insulin-like growth factor (IGF)-1, and IGF-binding protein 3 are described. |
| **Primary tissue or gland affected** | Adipocytes (leptin deficiency), hypothalamus (leptin receptor defect). |
| **Other organs, tissues, or glands affected** | Pituitary, gonads (hypogonadotropic hypogonadism), adrenals, bones (epiphyses). |
| **Age of onset** | Infancy or early childhood. |
| **Epidemiology** | Leptin disorders are extremely rare causes of obesity. Two kindreds with congenital leptin deficiency have been described: two children were affected in a Pakistani pedigree, and three siblings in a Turkish kindred. Three sisters with leptin receptor gene mutations have been described in a consanguineous family of Kabilian origin. |

| Inheritance | Autosomal recessive. |
|---|---|
| Chromosomal location | 7q31.3 (leptin), 1p31 (leptin receptor). |
| Genes | *LEP* (leptin), *LEPR* (leptin receptor). |
| Mutational spectrum | *LEP* gene mutations include a frame-shift mutation resulting from the deletion of a single guanine nucleotide in codon 133 of *LEP* and a missense mutation. A nucleotide base substitution in the splice donor site of exon 16 results in a leptin receptor defect. |
| Effect of mutation | The homozygous loss of function mutations of *LEP* result in congenital leptin deficiency. The homozygous mutations of *LEPR* result in a truncated receptor lacking both the transmembrane and intracellular domains. |
| Diagnosis | Measurement of serum leptin levels should be considered during the evaluation of severely obese patients with clinical features consistent with leptin deficiency or resistance. |
| Counseling issues | Families should be advised of the autosomal recessive pattern of inheritance for these disorders. Prenatal diagnosis may be possible when a specific mutation has been identified in an affected child. Treatment with recombinant human leptin may be successful in patients with congenital leptin deficiency. |
| Notes | Leptin is a 146-amino acid, cytokine-like peptide that is synthesized in adipocytes and binds to specific hypothalamic receptors. Leptin leads to a decrease in body weight by depressing appetite and increasing metabolic activity. In one patient with congenital leptin deficiency, daily injections of recombinant human leptin for 1 year resulted in weight loss, decreased food intake, increased energy expenditure, and pulsatile gonadotropin secretion. |

**References**

Montague CT, Farooqi IS, Whitehead JP, et al. Congenital leptin deficiency is associated with severe early-onset obesity in humans. *Nature* 1997;387:903–8.

Clement K, Vaisse C, Lahlou N, et al. A mutation in the human leptin receptor gene causes obesity and pituitary dysfunction. *Nature* 1998;392:398–401.

# 3. Disorders of the Thyroid Gland

# Primary disorders of thyroid gland function

Thyroid hormone synthesis involves a number of critical and unique steps, defects in any of which will cause defective thyroid hormone function. **Figure 1** demonstrates these major steps. The thyroid cell is stimulated by the action of thyrotropin (TSH) on a specific membrane-bound receptor (TSH receptor). Iodine is transported into the cell against a concentration gradient in an energy-dependent manner via the sodium–iodide symporter. Iodide is rapidly organified by a complex process requiring the recently identified iodide–chloride transporter pendrin and the enzyme thyroid peroxidase, both of which are present on the apical cell membrane. Successful organification requires these two proteins, an adequate source of $H_2O_2$, and the appropriate substrate, thyroglobulin (TG).

The iodinated tyrosine residues (monoiodotyrosine [MIT] and di-iodotyrosine [DIT]) are coupled to form tri-iodothyronine (T3) and thyroxine (T4). The hormone-bearing TG is endocytosed and hydrolyzed to form free T3 and T4, which are then released from the cell. MIT and DIT are di-iodinated and the resulting iodide is recycled by the cell.

A genetic defect in any of these steps will cause thyroid hormone dysgenesis, which presents clinically with variable degrees of hypothyroidism and goiter.

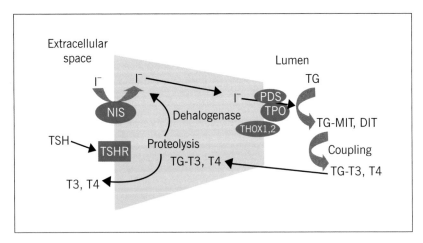

**Figure 1.** The major steps of thyroid hormone synthesis. DIT: di-iodotyrosine; MIT: monoiodotyrosine; NIS: sodium–iodide symporter; PDS: pendrin; T3: tri-iodothyronine; T4: thyroxine; THOX: thyroid oxidase (mutations in THOX2 have been shown to cause neonatal hypothyroidism); TG: thyroglobulin; TPO: thyroid peroxidase; TSH: thyrotropin; TSHR: TSH receptor.

*Genetics for Endocrinologists*

# Malignant tumors of the thyroid

Thyroid tumors can be divided into three major groups: differentiated tumors (papillary, follicular, and Hurthle cell tumors); medullary carcinoma, originating from the calcitonin-producing C cells of the thyroid; and anaplastic carcinoma. Familial forms of differentiated tumors and medullary carcinoma have been reported, and are discussed in this chapter. Anaplastic tumors occur sporadically; to date, no genetic basis has been defined.

# Autoimmune Thyroid Disease

(also known as: AITD; Hashimoto's thyroiditis; Hashimoto's struma; Graves' disease;
Basedow's disease)

| | |
|---|---|
| MIM | 275000 (Graves' disease; also called Basedow's disease), 140300 (Hashimoto's struma or thyroiditis) |
| Clinical features | AITD is a common endocrine disorder caused by the formation of autoantibodies to the thyrotropin (TSH) receptor, thyroglobulin, thyroid peroxidase (TPO), or other thyroid cell-specific proteins. Activating antibodies to the TSH receptor cause thyroid hyperfunction (Graves' disease), whereas antibodies to TPO are associated with destruction of the thyroid gland (Hashimoto's thyroiditis). Other antibodies, such as variants of receptor-activating and blocking antibodies, may cause hypofunction, goiter formation, or atrophy. Goiter is usually, though not invariably, present. When present, the typical Graves' disease goiter is soft and hypervascular, whereas a typical Hashimoto's goiter is firm and fibrotic. |
| Primary tissue or gland affected | Thyroid. |
| Other organs, tissues, or glands affected | Retro-orbital tissue, pretibial subcutaneous tissue. |
| Age of onset | Autoimmune thyroid disease can present in any age group, though it is rare before puberty and is most frequently seen between 20 years and 50 years of age. |
| Epidemiology | The prevalence of Graves' disease varies among populations, depending in part on iodine intake, and can be as high as 2% in women and 0.2% in men. The annual incidence of Hashimoto's thyroiditis is up to 4:1000 in women and 1:1000 in men. |
| Inheritance | Polygenic. |

| | |
|---|---|
| **Chromosomal location** | 6p21.3 (HLA), 2q33 (CTLA4), 14q31 (TSHR), 5q31-q33, 8q23-q24, 20q11.2 |
| **Genes** | *HLA* (human leukocyte antigen), *CTLA4* (cytotoxic T lymphocyte-associated antigen 4), *TSHR* (TSH receptor). |
| **Mutational spectrum** | AITD is a complex, polygenic disease. Familial clustering of AITD has long been recognized, though in most families inheritance appears to be complex, involving multiple genes and variable penetrance. Whole genome scans have suggested linkage to several different loci, though, for most, no candidate genes or specific mutations have been identified. Since most of the loci identified by whole genome scans have not been confirmed, it is likely that there are other loci that influence genetic susceptibility to AITD. The 6q21.3 and 5q31-q33 loci are linked to both forms of AITD (i.e. Graves' disease and Hashimoto's thyroiditis) in different populations, whereas 8q23-q24 is associated only with the latter. The allele frequencies of the major histocompatibility complex class 2 antigens HLA-DR3 and HLA-DQA1*0501 are increased in Graves' disease. Individuals with these haplotypes have a 2- to 5-fold greater risk of developing Graves' disease than those without. Polymorphisms in the *CTLA4* gene have also been associated with Graves' disease in several different populations, though it is not clear which, if any, of the polymorphisms studied have functional significance. CTLA4 polymorphisms are thought to confer as much as 50% of the inherited susceptibility in the populations studied. This same locus may also be associated with an increased risk of type 1 diabetes mellitus and Hashimoto's thyroiditis, perhaps explaining in part the association between type 1 diabetes mellitus and AITD. |
| **Effect of mutation** | Unknown. |
| **Diagnosis** | Hashimoto's thyroiditis is diagnosed by the presence of antithyroglobulin or anti-TPO antibodies, usually accompanied by |

thyroid dysfunction and goiter. Graves' disease is diagnosed by the presence of hyperthyroidism with suppressed TSH levels and diffuse hyperfunction of the gland. Most patients will have elevated levels of thyroid-stimulating antibody, the prevalence of which is largely dependent on the sensitivity of the assay used. Antithyroglobulin or anti-TPO antibodies may be present, but titers are usually lower than those seen in Hashimoto's thyroiditis. Associated autoimmune findings such as endocrine ophthalmopathy and pretibial myxedema can be seen in patients with either Graves' disease or Hashimoto's thyroiditis, but are much more common in those with Graves' disease.

**Counseling issues**

Families who demonstrate clear Mendelian inheritance of AITD are rare, though disease clustering within families is very common. Routine screening for AITD is not indicated, except in rare families with Mendelian inheritance of disease, though patients with a positive family history should warrant a high index of suspicion in order to avoid delays in diagnosis and treatment.

**Notes**

As with other complex genetic diseases, the interplay between genetic and environmental factors is still poorly understood. With the recent, rapid technological advances in human genetics, it may soon be possible to more clearly define the genetic risk factors associated with these diseases, and to correlate them with environmental risk factors. Improved understanding of the risk factors associated with specific diseases may provide the information needed to develop novel preventative and therapeutic interventions.

**References**

Barbesino G, Tomer Y, Concepcion ES et al. Linkage analysis of candidate genes in autoimmune thyroid disease II. Selected gender-related genes and the X-chromosome. International Consortium for the Genetics of Autoimmune Thyroid Disease. *J Clin Endocrinol Metab* 1998;83:3290–5.

Chistyakov DA, Savost'anov KV, Turakulov RI et al. Genetic determinants of Graves disease. *Mol Genet Metab* 2000;71:66–9 (Review).

*Autoimmune Thyroid Disease*

Sakai K, Shirasawa S, Ishikawa N et al. Identification of susceptibility loci for autoimmune thyroid disease to 5q31-q33 and Hashimoto's thyroiditis to 8q23-q24 by multipoint affected sib-pair linkage analysis in Japanese. *Hum Mol Genet* 2001;10:1379–86.

Yung E, Cheng PS, Fok TF et al. CTLA-4 gene A–G polymorphism and childhood Graves' disease. *Clin Endocrinol (Oxf)* 2002;56:649–53.

# Thyroid Dysgenesis

(includes: congenital hypothyroidism; thyroid agenesis; thyroid hypoplasia; ectopic thyroid)

| | |
|---|---|
| **MIM** | 218700, 241850 |

**Clinical features**  Typical clinical features of infants with untreated congenital hypothyroidism include lethargy, poor feeding, hypotonia, macroglossia, dry skin, hypothermia, hoarse cry, constipation, umbilical hernia, large fontanelle, and prolonged jaundice. Failure to initiate thyroid hormone replacement therapy promptly can result in severe and irreversible neurologic impairment. In two siblings with thyroid agenesis due to a mutation in the gene encoding thyroid transcription factor 2, hypothyroidism was associated with a cleft palate, bilateral choanal atresia, bifid epiglottis, and spiky hair. The prevalence of extrathyroidal anomalies, particularly cardiac defects (atrial and ventricular septal defects), ranges from 5.2% to 9% in children with congenital hypothyroidism, compared with prevalence rates of 2.5%–2.9% in the general population.

**Primary tissue or gland affected**  Thyroid.

**Age of onset**  Neonates and infants.

**Epidemiology**  Congenital hypothyroidism occurs in 1:3000–4000 live births. Thyroid dysgenesis accounts for 85% of all cases of congenital hypothyroidism, and approximately 2% of these patients have a positive family history for this disorder. An ectopic thyroid is found in two thirds, and hypoplasia or agenesis of the thyroid is present in one third, of both sporadic and familial forms of congenital hypothyroidism due to thyroid dysgenesis. Female predominance is noted for thyroid ectopy, but not for agenesis.

**Inheritance**  Autosomal recessive, autosomal dominant.

| Chromosomal location | 14q31 (TSHR), 9q22 (FKHL15), 2q12-q14 (PAX8) |
|---|---|
| Genes | *TSHR* (thyrotropin [TSH] receptor), *FKHL15* (forkhead-domain protein; also known as *TTF2* [thyroid transcription factor 2]), *PAX8* (paired box 8). |
| Mutational spectrum | Nucleotide substitutions, deletions, and splicing mutations (*TSHR*); nucleotide substitutions (*FKHL15* and *PAX8*). |
| Effect of mutation | Loss of function mutation of *TSHR* results in a hypoplastic, hypofunctioning, but normally located thyroid gland. Substitutions described for *FKHL15* result in missense mutations and markedly decreased transcription factor, leading to thyroid agenesis. Mutant *PAX8* is unable to activate a reporter gene under the control of the human thyroid peroxidase promoter. The *PAX8* mutation interferes with the early stages of thyroid development, and results in thyroid agenesis or hypoplasia. |
| Diagnosis | Because typical signs and symptoms of congenital hypothyroidism are nonspecific and often absent early on in affected infants, routine neonatal screening is essential to arrive at the diagnosis and initiate treatment promptly. When congenital hypothyroidism is confirmed by low free T4 and elevated TSH levels (compared with age-matched reference values), radioisotope scanning can identify the specific type of thyroid gland abnormality. Serum thyroglobulin levels may be helpful in distinguishing thyroid agenesis from other causes of an 'invisible' thyroid, such as the presence of blocking antibodies. |
| Counseling issues | When counseling families of infants with congenital hypothyroidism, the 2% prevalence of familial cases should be noted along with the small, but significantly increased, risk of extrathyroidal anomalies. |
| Notes | The clinical sydrome of congenital hypothyroidism due to aplasia, hypoplasia, or ectopic thyroid tissue can be caused by mutations in the different genes responsible for thyrogenesis. TSHR mutations |

causing thyroid dysgenesis are rare; more commonly, these mutations result in variable degrees of resistance to TSH in patients with an anatomically normal thyroid gland.

**References**

Castanet M, Polak M, Bonaiti-Pellie C et al. Nineteen years of national screening for congenital hypothyroidism: Familial cases with thyroid dysgenesis suggest the involvement of genetic factors. *J Clin Endocrinol Metab* 2001;86:2009–14.

Vilain C, Rydlewski C, Duprez L et al. Autosomal dominant transmission of congenital thyroid hypoplasia due to loss-of-function mutation of PAX8. *J Clin Endocrinol Metab* 2001;86:234–8.

# Thyroid Hormonogenesis Defect 1

**(also known as: iodine transport or trapping defect)**

| | |
|---|---|
| **MIM** | 274400 |
| **Clinical features** | The clinical hallmark of this disease is multinodular goiter, with or without clinical hypothyroidism, which appears during childhood. Consanguinity is a common finding, given that this is a very rare recessive disorder. Although some patients remain euthyroid, most have some degree of hypothyroidism, the severity of which appears to be related to iodine intake. Patients from Greece and Brazil, areas of iodine deficiency, tend to be more severely hypothyroid, with irreversible mental retardation if not diagnosed and treated early. Dietary iodine supplementation can restore euthyroidism, though thyroid hormone replacement appears to be a wiser therapeutic alternative. |
| **Primary tissue or gland affected** | Thyroid. |
| **Age of onset** | Goiter usually develops in early childhood. A severe defect, along with dietary iodine deficiency, will result in congenital hypothyroidism. |
| **Epidemiology** | Judging from the small number of cases reported, this appears to be one of the rarest of the genetic disorders of thyroid function. |
| **Inheritance** | Autosomal recessive. |
| **Chromosomal location** | 19p13.2-p12 |
| **Genes** | *SLC5A5* (sodium–iodide symporter; also known as NIS). |
| **Mutational spectrum** | A single point mutation in codon 354 (Thr354Pro) and a nonsense mutation (C272X) have been described, both of which cause severe |

defects in transporter function. Missense mutations may also exist, causing milder defects.

**Effect of mutation**     Without a functional sodium–iodide symporter, the thyroid cell is unable to maintain a concentration gradient of readily exchangeable iodine between the plasma and the cytoplasm. The result is insufficient intracellular iodide for thyroid hormonogenesis.

**Diagnosis**     The clinical diagnosis is based on the demonstration of low iodine uptake accompanied by elevated thyrotropin levels. Other causes of low iodine uptake, such as prior administration of large doses of iodine, must be excluded. Decreased or absent transport of iodine into the salivary glands can also be demonstrated. Thyroid tissue iodine content is low, and inorganic iodine supplementation will correct the hypothyroidism. A definitive diagnosis is made by identifying a mutation in the sodium–iodide symporter gene.

**Counseling issues**     Early diagnosis and treatment will prevent hypothyroidism, mental retardation, growth retardation, and goiter.

**References**     Fujiwara H, Tatsumi K, Miki K et al. Congenital hypothyroidism caused by a mutation in the Na+/I⁻ symporter. *Nat Genet* 1997;16:124–5.

# Thyroid Hormone Organification Defect 2

(also known as: iodide peroxidase deficiency; thyroid peroxidase deficiency)

| | |
|---|---|
| **MIM** | 274500, 606759 |

**Clinical features**   Complete functional defects of thyroid peroxidase (TPO) result in severe congenital hypothyroidism or cretinism. Stimulated by high levels of thyrotropin, the thyroid gland grows and typically develops into a large, multinodular goiter during early or mid-childhood. Hearing is not affected, and the coexistence of an iodide organification defect and sensorineural hearing loss suggests the presence of Pendred syndrome (MIM 274600, next entry), which is caused by mutations in a different protein thought to function as an iodide transporter at the apical border of the thyroid cell. Milder functional defects of TPO will result in less severe hypothyroidism. Some patients have normal thyroid function, but develop recurrent goiters during childhood.

**Primary tissue or gland affected**   Thyroid.

**Age of onset**   Birth.

**Epidemiology**   Congenital hypothyroidism due to a complete organification defect was reported in 1 in 66 000 live births in The Netherlands.

**Inheritance**   Autosomal recessive.

**Chromosomal location**   2p25 (TPO), 15p15.3 (THOX2)

**Genes**   *TPO* (thyroid peroxidase), *THOX2* (thyroid oxidase).

**Mutational spectrum**   Missense, termination, insertion, and duplication mutations have been described. The insertion and duplication mutations typically cause frame shifts and premature termination. Recently, inactivating mutations in the *THOX2* gene, one of the components of the hydrogen peroxide-generating system in the thyroid, were shown to cause a defect in thyroid hormonogenesis.

**Effect of mutation**

*TPO* mutations result in decreased enzyme function. Premature termination, insertion, and duplication mutations result in a total lack of functional enzyme, causing severe disease. *THOX2* mutations result in decreased hydrogen peroxide production, which leads to a secondary decrease in TPO activity. *THOX2* mutations have been described in patients with mild or severe organification defects, clinically indistinguishable from those with severe *TPO* mutations. Partial iodide-organification defects and transient congenital hypothyroidism have been found in patients heterozygous for *THOX2* mutations, while severe congenital hypothyroidism has been found in a patient homozygous for a similar mutation.

**Diagnosis**

Clinical diagnosis is made by demonstrating poor organification of a radioactive iodine tracer. This can be documented by performing a perchlorate discharge test. Perchlorate blocks iodide uptake, but permits leakage of nonorganified intracellular iodide out of the cell. Discharge of >10% of labeled iodide from the thyroid is considered pathologic. Other causes of impaired organification – including Pendred syndrome and thyroglobulin defects – must be excluded.

**Counseling issues**

As with other forms of congenital hypothyroidism, early diagnosis and treatment can prevent the mental and growth retardation associated with this syndrome.

**Notes**

Defects in other factors required for normal TPO enzymatic function may cause a clinically similar syndrome. Specifically, a defective prosthetic group required for enzyme function will result in a similar organification defect.

**References**

Ambrugger P, Stoeva I, Biebermann H et al. Novel mutations of the thyroid peroxidase gene in patients with permanent congenital hypothyroidism. *Eur J Endocrinol* 2001;145:19–24.

*Thyroid Hormone Organification Defect 2*

Bakker B, Bikker H, Vulsma T et al. Two decades of screening for congenital hypothyroidism in The Netherlands: TPO gene mutations in total iodide organification defects (an update). *J Clin Endocrinol Metab* 2000;85:3708–12.

Krude H, Biebermann H, Schnabel D. Molecular pathogenesis of neonatal hypothyroidism. *Horm Res* 2000;53(Suppl. 1):12–8 (Review).

Moreno JC, Bikker H, Kempers MJ et al. Inactivating mutations in the gene for thyroid oxidase 2 (THOX2) and congenital hypothyroidism. *N Engl J Med* 2002;347:95–102.

# Pendred Syndrome

(also known as: PDS; goiter–deafness syndrome; thyroid hormonogenesis defect 2B)

**MIM**          274600

**Clinical features**          The hallmark of PDS is the coexistence of severe sensorineural deafness with goiter. Sensorineural deafness is usually congenital (prelingual). However, in some families deafness appears later in life, usually during childhood, after some mild metabolic insult such as viral infection or surgery. Thyroid function may be normal, or patients may have subclinical hypothyroidism with mildly elevated thyrotropin (TSH), but normal levels of tri-iodothyronine and thyroxine. TSH response to thyroid-releasing hormone stimulation is typically exaggerated. Thyroglobulin levels are typically very elevated. Frank hypothyroidism has been reported in rare cases. Goiter typically develops during childhood or adolescence, and thus the diagnosis may be missed at the onset of deafness. The goiter may become very large, causing compression symptoms (see **Figure 2**). Papillary carcinoma has been reported. Several anatomical defects have been described in the inner ear, including Mondini dysplasia of the cochlea; however, widening of the vestibular aqueduct appears to be the most common and consistent finding (see **Figure 3**). Mutations in *SLC26A4* (pendrin) may also cause nonsyndromic hereditary sensorineural deafness (DFNB4). Mutations in this gene may be responsible for 1%–10% of all hereditary deafness.

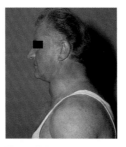

**Figure 2.** Multinodular goiter such as is seen in Pendred syndrome.

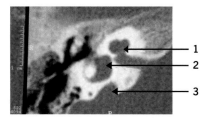

**Figure 3.** CT image of an ear in a patient with Pendred syndrome. The numbers indicate: (1) a mildly deformed cochlea; (2) enlarged, ballooned vestibular complex extending into the semicircular canals; and (3) a widened vestibular aqueduct.

| | |
|---|---|
| **Primary tissue or gland affected** | Thyroid. |
| **Other organs, tissues, or glands affected** | Ear. |
| **Age of onset** | Congenital (see clinical features). |
| **Epidemiology** | Very rare, except in consanguineous populations. An incidence of less than 1 in 10 000 has been reported in the British Isles. In contrast, a single Bedouin tribe with more than 35 affected members has been reported in Israel. |
| **Inheritance** | Autosomal recessive. |
| **Chromosomal location** | 7q31 |
| **Genes** | *SLC26A4* (pendrin). |
| **Mutational spectrum** | Missense, deletion, and splice mutations have been reported. |
| **Effect of mutation** | The mutations result in lack of production of functional protein. The defect may be at the level of transcription, translation (stop codons), protein processing, translocation to the membrane, or translocation function. |

**Diagnosis**

Diagnosis is based on the finding of sensorineural deafness in combination with mild iodine organification defect. The latter can be demonstrated by performing a potassium perchlorate discharge test; however, this test is not specific and its sensitivity is unknown. TSH may be somewhat elevated, but frank hypothyroidism is not common. Thyroglobulin levels may be extremely high, even in the absence of any evidence of malignancy. Definitive diagnosis requires the identification of a functional mutation in the *SLC26A4* gene.

**Counseling issues**

Mutation analysis will permit genetic counseling and prenatal diagnosis. Early treatment with thyroid hormone may prevent the development of goiter, though this has not been proven.

**Notes**

Mutations in the *SLC26A4* gene should be considered in all cases of sensorineural deafness, even in the absence of thyroid abnormalities.

**References**

Everett LA, Glaser B, Beck JC et al. Pendred syndrome is caused by mutations in a putative sulphate transporter gene (*PDS*). *Nat Genet* 1997;17:411–22.

Phelps PD, Coffey RA, Trembath RC et al. Radiological malformations of the ear in Pendred syndrome. *Clin Radiol* 1998;53:268–73.

# Iodothyrosine Dehalogenase Deficiency

(also known as: thyroid hormonogenesis defect 4)

| | |
|---|---|
| MIM | 274800 |

**Clinical features**
Patients with severe iodothyrosine dehalogenase defects have congenital hypothyroidism and cretinism, which is caused by iodine depletion, and not defective hormonogenesis *per se*. If dietary iodine is sufficient, hormone production may be normal. Goiter may be present at birth or develop during childhood.

**Primary tissue or gland affected**
Thyroid.

**Age of onset**
Birth.

**Epidemiology**
The syndrome is extremely rare, and was first described in the inbred tinkers, a nomadic sect in western Scotland. Subsequently, other isolated groups have been reported. Mild deiodinase defects may be much more common, and may go undetected if dietary iodine is sufficient to correct for the increased iodine loss.

**Inheritance**
Autosomal recessive.

**Chromosomal location**
Unknown.

**Genes**
Specific genetic mutations have not yet been identified.

**Mutational spectrum**
Unknown.

**Effect of mutation**
Patients with dehalogenase defects lack the ability to deiodinate monoiodotyrosine (MIT) and di-iodotyrosine (DIT), resulting in insufficient intracellular substrate for thyroid hormone production, and in urinary iodine loss.

**Diagnosis**

The syndrome should be suspected in patients with early onset of goiter and hypothyroidism. Initial iodine uptake is very high, and a large fraction of the accumulated labeled iodine is lost from the gland within a few hours. High serum and urine levels of MIT and DIT are diagnostic of a thyroid deiodinase disorder. Urinary excretion of radiolabeled MIT and DIT can be measured easily after oral administration of a tracer dose of radioactive iodine. A generalized lack of deiodinase can be diagnosed by administering radiolabeled DIT and measuring urinary secretion of the unmetabolized compound.

**Counseling issues**

As with other defects of thyroid hormonogenesis, early diagnosis and adequate thyroxine replacement will prevent all symptoms related to hypothyroidism. Thus, this treatment is much more efficient and effective than the seemingly more physiologic treatment with dietary iodine supplementation.

**References**

Krude H, Biebermann H, Schnabel D. Molecular pathogenesis of neonatal hypothyroidism. *Horm Res* 2000;53:12–8.

# Thyroglobulin Gene Mutations

| | |
|---|---|
| MIM | 188450 |
| Clinical features | Patients may present with congenital hypothyroidism with goiter, or goiter may appear during childhood or adolescence. The clinical presentation is similar to other forms of defective thyroid hormonogenesis. |
| Primary tissue or gland affected | Thyroid. |
| Age of onset | Birth. |
| Epidemiology | This is a very rare disorder. Differential diagnosis between this and other forms of thyroid dyshormonogenesis can be difficult, and the actual prevalence may be greater than suspected. |
| Inheritance | Autosomal recessive, autosomal dominant. |
| Chromosomal location | 8q24.2-q24.3 |
| Genes | *TG* (thyroglobulin). |
| Mutational spectrum | Missense, splicing, termination, and deletion mutations have all been identified. |
| Effect of mutation | Mutations result in failure to synthesize TG, abnormal mRNA or protein processing, or in the synthesis of abnormal protein. Lack of protein or abnormal protein will prevent or reduce thyroid hormone production, thus causing a recessive defect. Alternatively, an abnormal protein may form a complex with a normal molecule, thereby interfering with hormonogenesis and producing a dominant negative effect. |

**Diagnosis**

Differential diagnosis between this and other forms of thyroid hormone dysgenesis may be difficult since TG may be measurable by immunoassay, but is biologically inactive. In one family, a splice-site mutation resulted in production of a truncated protein lacking exon 4, which contains a putative donor tyrosine residue. The result was a slightly smaller than normal protein, with very limited biological activity. In some patients, TG levels are markedly decreased, confirming the diagnosis.

**Counseling issues**

As with other forms of thyroid dyshormonogenesis, early diagnosis and effective treatment with exogenous thyroid hormone prevents the debilitating effects of hypothyroidism.

# Thyroid Hormone Resistance

(also known as: pituitary, peripheral, generalized, or combined thyroid hormone resistance)

| | |
|---|---|
| **MIM** | 188570 |
| **Clinical features** | Affected patients typically show few, if any, signs or symptoms of thyroid dysfunction, despite elevated concentrations of serum tri-iodothyronine (T3) and thyroxine (T4). The most frequent manifestations are goiter, attention-deficit hyperactivity disorder, and tachycardia. Growth retardation, delayed bone age, emotional disturbances, learning disabilities, mental retardation, and hearing loss are common. Impaired hearing may be due to an increased frequency of recurrent ear infections, and to sensorineural deafness, which is associated with thyroid hormone receptor (THR)-β gene deletions. |
| **Primary tissue or gland affected** | Thyroid. |
| **Other organs, tissues, or glands affected** | Central nervous system, heart, ears, bones |
| **Age of onset** | Clinical manifestations of thyroid hormone resistance typically present in childhood, but the diagnosis may be made at any age during the evaluation of elevated serum T3 and T4 levels associated with normal or high thyrotropin (TSH) concentrations. |
| **Epidemiology** | The estimated prevalence of resistance to thyroid hormone is 1:50 000 live births. |
| **Inheritance** | Autosomal dominant, autosomal recessive. |
| **Chromosomal location** | 3q24.3 |
| **Genes** | *THRB* (thyroid hormone receptor-β). |

**Mutational spectrum**   A total of 101 mutations have been identified in 167 unrelated families. Of these, 15% are *de novo* mutations. Mutant gene products interfere with the function of normal receptors (dominant negative mutations). Most mutations are clustered in the T3-binding domain (exons 8–10). Deletions and substitutions have been described. Deletion of the *THRB* gene is present in the recessive form. Some families with thyroid hormone resistance have no detectable mutations in the *THRB* gene; mutations in cofactors that modulate receptor function may explain the hormone resistance observed in these families.

**Effect of mutation**   Mutant *THRB* blocks the transcriptional activity of wild-type *THR* (dominant negative activity), possibly by the formation of inactive wild-type/mutant THR dimers. Mutant THR may release corepressors that contribute to the dominant negative effect. Resistance to thyroid hormone at the pituitary level results in increased TSH secretion, which stimulates the production of T3 and T4. The elevated thyroid hormone levels compensate, to variable extents, for the peripheral tissue resistance to T3 and T4.

**Diagnosis**   Documentation of persistently elevated concentrations of serum free T3 and T4, together with normal or slightly increased TSH levels, is highly suggestive of resistance to thyroid hormones. These laboratory findings may also be present in patients with a TSH-producing adenoma of the pituitary. However, pituitary adenoma patients are usually clinically hyperthyroid. The presence of first-degree relatives with similar laboratory findings makes thyroid hormone resistance more likely than pituitary adenoma. Magnetic resonance imaging of the pituitary and measurement of serum glycoprotein $\alpha$-subunit levels will assist in differentiating an adenoma from thyroid hormone resistance. Serum levels of cholesterol, carotene, ferritin, and sex hormone-binding globulin, along with measurement of the pulse rate and basal metabolic rate, may be helpful in assessing peripheral thyroid hormone resistance. Monitoring TSH levels during a T3 suppression test demonstrates the degree of pituitary resistance to thyroid hormones.

**Counseling issues**     Awareness of thyroid hormone resistance in affected families is important in order to prevent unwarranted antithyroid treatment of patients with elevated serum T3 and T4 levels, and to distinguish patients from those with a TSH-secreting pituitary adenoma.

**Notes**     The degree of resistance in peripheral tissue is sometimes greater than the pituitary receptor defect. When this occurs, it may be necessary to raise serum T4 levels by administering exogenous L-thyroxine in order to compensate for the hormone resistance. Since TSH cannot be used as a marker for adjusting L-thyroxine dosage, other parameters, such as growth, skeletal maturation, and neurological development in children – in addition to monitoring basal metabolic rate, nitrogen balance, and circulating sex hormone-binding globulin levels – may be used to titrate the L-thyroxine dosage.

**References**     Brucker-Davis F, Skarulis MC, Grace MB. Genetic and clinical features of 42 kindreds with resistance to thyroid hormone. The National Institutes of Health Prospective Study. *Ann Intern Med* 1995;123:572–83.

Phillips SA, Rotman-Pikielny P, Lazar J et al. Extreme thyroid hormone resistance in a patient with a novel truncated TR mutant. *J Clin Endocrinol Metab* 2001;86:5142–7.

# Thyrotropin Resistance

(also known as: unresponsiveness to thyrotropin [TSH]; congenital hypothyroidism due to TSH resistance)

| | |
|---|---|
| **MIM** | 275200 |
| **Clinical features** | The variation in phenotypes of affected patients represents the different degrees of impaired TSH receptor function. Partial impairment of the receptor results in normal to low serum thyroxine (T4) levels, with elevated TSH concentrations (compensated or subclinical hypothyroidism). Complete absence of receptor function is associated with severe hypothyroidism, and, in some cases, a markedly hypoplastic thyroid gland. |
| **Primary tissue or gland affected** | Thyroid. |
| **Age of onset** | Usually diagnosed at birth during routine neonatal screening for congenital hypothyroidism. |
| **Epidemiology** | Resistance to TSH is a rare cause of congenital hypothyroidism. Only 15 loss of function mutations in 11 pedigrees have been reported. |
| **Inheritance** | Autosomal recessive. |
| **Chromosomal location** | 14q31 |
| **Genes** | *TSHR* (thyrotropin receptor). |
| **Mutational spectrum** | Fifteen loss of function mutations (substitutions and deletions) of the *TSHR* gene have been described, of which nine are in the extracellular domain and six in the transmembrane segment. In some patients with TSH resistance, no TSHR mutation can be identified – in some of these, linkage analysis has excluded the TSHR locus, strongly suggesting that other genetic defects can cause the same clinical phenotype. |

| | |
|---|---|
| **Effect of mutation** | The mutations affect ligand binding, signal transduction, and/or cell surface expression of the receptor. |
| **Diagnosis** | The presence of a normally located thyroid gland in infants, with high serum TSH and normal to low levels of free tri-iodothyronine and T4, suggests the possibility of TSH resistance. An autosomal recessive pattern of inheritance, a normally located, nongoitrous thyroid with low radioactive iodine uptake, absent *in vivo* response to exogenous TSH, and an absence of antithyroid antibodies add further support for this diagnosis. Resistance to TSH is unlikely when a goiter or ectopic thyroid tissue are present. |
| **Counseling issues** | In contrast to the sporadic nature of most cases of congenital hypothyroidism, the autosomal recessive inheritance of TSH resistance may lead to early diagnosis and initiation of L-thyroxine treatment in the offspring of affected families. |
| **Notes** | As with other causes of congenital hypothyroidism, prompt initiation of L-thyroxine treatment is essential in order to avoid irreversible neurologic and developmental damage. |
| **References** | Clifton-Bligh RJ, Gregory JW, Ludgate M et al. Two novel mutations in the thyrotropin (TSH) receptor gene in a child with resistance to TSH. *J Clin Endocrinol Metab* 1997;82:1094–100. |
| | Nagashima T, Murakami M, Onigata K et al. Novel inactivating missense mutations in the thyrotropin receptor gene in Japanese children with resistance to thyrotropin. *Thyroid* 2001;11:551–9. |

# Thyroxine-binding Globulin Deficiency

**(also known as: TBG deficiency; complete [TBG-CD] or partial TBG deficiency)**

| | |
|---|---|
| MIM | 314200 |

**Clinical features**  Total thyroxine and tri-iodothyronine (T3) levels are low, though free T3, free thyroxine (T4), and thyrotropin serum levels are normal. Individuals with TBG deficiency are euthyroid, and this anomaly has not been associated with clinical disease.

**Primary tissue or gland affected**  Thyroid.

**Age of onset**  The defect is present from birth, but is not associated with clinical disease, and thus is not diagnosed until thyroid hormone levels are tested.

**Epidemiology**  Reported prevalence varies from 1:2800–15 000 males. Milder forms of TBG deficiency may occur in 1:42 000 heterozygous females.

**Inheritance**  X-linked.

**Chromosomal location**  Xq22.2

**Genes**  *TBG* (thyroxin-binding globulin).

**Mutational spectrum**  Single nucleotide substitutions, deletions, additions, and missense mutations have been described.

**Effect of mutation**  Mutations that cause an early stop codon produce a truncated TBG molecule. Substitution of the normal Leu227 with a proline in TBG-CD5 causes failure of secretion of the variant molecule because of aberrant posttranslational processing.

**Diagnosis**

A finding of low serum total T4 in the presence of normal free T4 and thyrotropin strongly suggests the diagnosis. A direct assay of TBG shows low levels in affected males (or in affected 45,X females), and intermediate levels in female carriers.

**Counseling issues**

Affected individuals and their families need to understand the benign nature of this condition ('nondisease'), and to avoid inappropriate treatment in the event that a low serum total T4 level is misinterpreted as evidence for hypothyroidism.

**Notes**

Recent studies have determined that TBG is a member of the serpin (serine protease inhibitor) family of proteins. Other members of this family include antithrombin, $\alpha$-1-antitrypsin, and plasminogen activator inhibitors. A distinctive property of this family of proteins is the structural change that occurs in response to limited proteolysis. Exposure of TBG to pancreatic elastase or activated polymorphonuclear leukocytes results in a major decrease in its affinity for T4. A possible role for TBG-mediated release of T4 in response to inflammation or other physiological processes has been suggested.

**References**

Jirasakuldech B, Schussler GC, Yap MG et al. A characteristic serpin cleavage product of thyroxine-binding globulin appears in sepsis sera. *J Clin Endocrinol Metab* 2000;85:3996–9.

Schussler GC. The thyroxine-binding proteins. *Thyroid* 2000;10:141–9 (Review).

# Familial Dysalbuminemic Hyperthyroxinemia

(also known as: FDH)

| | |
|---|---|
| **MIM** | 103600 |
| **Clinical features** | Free tri-iodothyronine (T3) and free thyroxine (T4) levels are normal, patients are euthyroid, and, except for complicating the interpretation of thyroid function tests, FDH causes no detrimental health effects. |
| **Primary tissue or gland affected** | Albumin production in liver; thyroid. |
| **Age of onset** | Although the disorder is present from birth, FDH may be diagnosed at any age when elevated total T3 or total T4 serum levels are noted. |
| **Epidemiology** | The estimated prevalence of FDH in a general population is approximately 1:100. |
| **Inheritance** | Autosomal dominant. |
| **Chromosomal location** | 4q11-q13 |
| **Genes** | *ALB* (human albumin gene). |
| **Mutational spectrum** | Three mutations causing FDH have been identified: R218H, R218P, and L66P. The R218H mutation is the most common cause of FDH in a white population. |
| **Effect of mutation** | Depending on the specific mutation, changes in affinity for T3 and T4 result in variable increases in serum concentrations of these hormones. R218H raises serum total T4 to levels 2–3 times above the normal range, while total T3 levels are normal or only slightly elevated. The R218P defect raises T4 11- to 17-fold above normal, but T3 levels are less elevated. In patients with the L66P mutation, serum T3 is elevated, but total T4 levels are normal. |

| | |
|---|---|
| **Diagnosis** | Elevated serum concentrations of total T3 and/or T4 together with normal free T3, T4, and thyrotropin levels in a clinically euthyroid patient indicate an excess of thyroid hormone-binding proteins. Similar findings occur in congenital or acquired thyroxine-binding globulin (TBG) excess, which can be excluded by measuring TBG levels. The initial differential diagnosis may include hyperthyroidism or thyroid hormone resistance. Although free T4 levels are normal when measured by equilibrium dialysis, levels may appear to be elevated when using some standard free T4 immunoassay kits. |
| **Counseling issues** | Families should be apprised of the autosomal dominant hereditary pattern, and that FDH is 'a condition and not a disease' in order to avoid unnecessary medical testing and/or inappropriate treatment. |
| **Notes** | Because of the relatively high prevalence of FDH, some patients with acquired or congenital hypothyroidism may also have FDH. The coexistence of these two entities leads to difficulties in diagnosis and in monitoring treatment. |
| **References** | Pohlenz, J, Sadow PM, Koffler T et al. Congenital hypothyroidism in a child with unsuspected familial dysalbuminemic hyperthyroxinemia caused by a mutation (R218H) in the human albumin gene. *J Pediatr* 2001;139:887–91. |

# Familial Nonmedullary Thyroid Cancer

(also known as: FNMTC)

| | |
|---|---|
| **MIM** | 188550, 605642, 603386 |
| **Clinical features** | FNMTC is characterized by a higher rate of multicentricity and more aggressive behavior than its nonfamilial counterpart. Several different chromosomal loci have been reported, and each appears to be associated with a slightly different clinical phenotype. The Ch1q21 locus is associated with coexistence of the common variety of papillary thyroid carcinoma and papillary renal neoplasia. Ch19p13.2 is associated with NMTC with oxyphilia. Both of these subtypes appear to be particularly rare. The 2q21 locus was identified in a large Tasmanian pedigree, and confirmed in a larger, ethnically diverse population. This locus appears to be associated with a high incidence of the follicular variant of papillary carcinoma. Families with both thyroid and colon cancer have also been described. |
| **Primary tissue or gland affected** | Thyroid. |
| **Other organs, tissues, or glands affected** | Families with coexistant colon, renal, and other abdominal cancers have been reported. |
| **Age of onset** | Tumors are usually identified during the fourth or fifth decade of life. |
| **Epidemiology** | The overall incidence of familial thyroid carcinoma is unknown. |
| **Inheritance** | Autosomal dominant. |
| **Chromosomal location** | 10q21 (RET), 1q21, 19p13.2, 2q21. |
| **Genes** | *RET* (ret proto-oncogene), unidentified genes at other loci. |

| | |
|---|---|
| **Mutational spectrum** | Most families with FNMTC are too small to permit linkage analysis. Multiple chimeric genes have been reported in which the tyrosine kinase domain of the *RET* proto-oncogene is joined to ubiquitously expressed activating genes. Several different chimeric genes have been reported in familial and sporadic tumors. |
| **Effect of mutation** | Chimeric recombinations result in activation of the tyrosine kinase segment of the *RET* proto-oncogene. |
| **Diagnosis** | At present, there are no specific clinical or genetic signs or symptoms that can differentiate familial from sporadic tumors. The association of papillary thyroid cancer with colon or renal cancer suggests the possibility of familial predisposition. |
| **Counseling issues** | Familial disease is usually inherited as an autosomal dominant trait with variable penetrance. |
| **Notes** | Since sporadic carcinoma cannot be differentiated from familial cancer on the basis of clinical or histologic findings, it is prudent to evaluate close family members of thyroid cancer patients with particular care. |
| **References** | McKay JD, Lesueur F, Jonard L et al. Localization of a susceptibility gene for familial nonmedullary thyroid carcinoma to chromosome 2q21. *Am J Hum Genet* 2001;69:440–6. |
| | Ruben Harach H. Familial nonmedullary thyroid neoplasia. *Endocr Pathol* 2001;12:97–112 (Review). |

# Familial Medullary Carcinoma of the Thyroid

**(also known as: FMTC)**

| | |
|---|---|
| **MIM** | 155240 |
| **Clinical features** | Patients may present with a thyroid mass or with symptoms of calcitonin over-secretion, such as diarrhea and flushing. As with other familial forms of MTC (multiple endocrine neoplasia [MEN]2a and 2b; see p. 6), this disease is often bilateral and is associated with C-cell hyperplasia. Unless diagnosed as part of a family screening program, the tumor has usually metastasized when discovered. |
| **Primary tissue or gland affected** | Thyroid. |
| **Age of onset** | The mean age of onset is 37 years. |
| **Epidemiology** | MTC accounts for approximately 5%–10% of all thyroid carcinomas, and about 25% of cases are familial. Of these, 15% are part of the MEN2a syndrome, 1% the MEN2b syndrome, and 9% are not associated with any other endocrinopathy or tumor (FMTC). |
| **Inheritance** | Autosomal dominant. |
| **Chromosomal location** | 10q11.2 |
| **Genes** | *RET* (ret proto-oncogene). |
| **Mutational spectrum** | Missense mutations have been identified in exons 8, 10, 11, 13, 14, and 15 of the *RET* proto-oncogene. Mutations can be identified in about 90% of FMTC families. |
| **Effect of mutation** | Mutations activate *RET* to function as a dominant oncogenic transforming gene. |

**Diagnosis**

The diagnosis of familial disease is suspected when bilateral disease or C-cell hyperplasia is discovered. Definitive diagnosis is made by screening the appropriate exons of the *RET* proto-oncogene. No mutations will be identified in exons 8, 10, 11, 13, 14, or 15 in 7%–10% of patients. If possible, linkage analysis should be perfomed in these patients to determine if association with the *RET* locus (10q11.2) can be confirmed.

**Counseling issues**

Treatment for metastatic MTC is palliative, and mortality, though frequently delayed, is very high. Even in the presence of metastatic disease, aggressive surgical and medical treatment may significantly prolong survival. All patients with MTC must be informed of the 25% chance of familial disease. If a mutation is identified, all family members should be screened. Siblings and children of the proband have a 50% chance of inheriting the mutation, and penetrance approaches 100%. Family members carrying a mutation should be tested for MTC and its precursor, C-cell hyperplasia. Total thyroidectomy should be performed if calcitonin levels are abnormal, or if there is an abnormal response to calcium–pentagastrin stimulation. Prophylactic total thyroidectomy should be considered, even if the calcitonin response is normal. Since the same mutations can cause both FMTC and MEN1a, and since pheochromocytoma may appear years after the diagnosis of MTC in MEN1a patients, all FMTC patients should be screened periodically for pheochromocytoma.

**Notes**

As with other familial forms of MTC, mutation analysis and family screening are critical, since the only possibility for cure is total thyroidectomy before the appearance of any extrathyroidal metastases. This is not usually possible if the disease is discovered on the basis of a thyroid mass or clinical symptoms.

**References**

Modigliani E, Franc B, Niccoli-Sire P. Diagnosis and treatment of medullary thyroid cancer. *Baillieres Best Pract Res Clin Endocrinol Metab* 2000;14:631–49 (Review).

# 4. Disorders of Calcium and Bone Metabolism

# Syndromes associated with hypercalcemia

This chapter begins with a discussion of genetic entities whose major clinical feature is hypercalcemia. Hypercalcemia is found in several genetic syndromes that are associated with other endocrine abnormalities, and are thus discussed in Chapter 1. Two of the best known are the multiple endocrine neoplasia (MEN) syndromes, MEN1 (MIM 131100) and MEN2 (MIM 171400, 162300).

# Syndromes associated with hypocalcemia

Next, we discuss those genetic entities whose major clinical feature is hypocalcemia. As with hypercalcemia, hypocalcemia may be found in several complex genetic syndromes, including autoimmune polyendocrinopathy type 1 (MIM 240300), and thalassemia major (MIM 141900), which are discussed in Chapter 1.

# Syndromes associated with abnormal vitamin D metabolism

Vitamin D functions as a hormone, whose activation is dependent on parathyroid hormone and whose cellular function is dependent on binding to a specific nuclear receptor (see **Figure 1**). Abnormalities in vitamin D metabolism and function are included in this section.

# Primary bone disorders

Although some of the entities presented in this section are not primary endocrine diseases, they are included in this chapter because their presenting symptoms frequently precipitate referral to the endocrinologist for further evaluation, diagnosis, and treatment. Hypophosphatasia may present at any age with varying severity, ranging from fetal death to mild dental abnormalities. Infantile hypercalcemia or childhood rickets may be the presenting symptoms that initiate an endocrine referral.

Osteoporosis is a very common entity, and increased public awareness has resulted in a marked increase in referrals. The common form of the disease described here is associated with the normal, age-related decrease in sex hormone production related to the menopause in women and advanced age in men. Similarly, osteogenesis imperfecta and skeletal dysplasias are not primary endocrine diseases, but they are frequently encountered in the evaluation of short stature, and familiarity with these entities is therefore required of the endocrinologist.

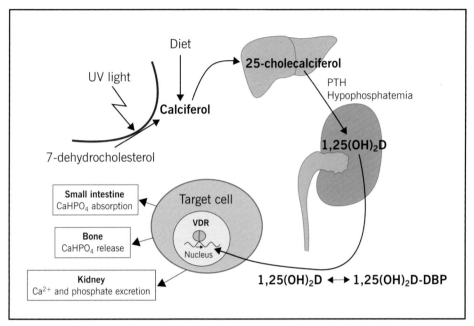

**Figure 1.** Vitamin D metabolism and action. Calciferol, the vitamin D prehormone is obtained either from the diet, or from nonenzymatic photolysis of 7-dehydrocholesterol (provitamin D) in response to near-ultraviolet (UV)-light radiation to the skin. Calciferol undergoes 25 hydroxylation in the liver. 25(OH)D then undergoes 1 $\alpha$-hydroxylation in the kidney, a reaction regulated by parathyroid hormone (PTH) and stimulated by hypophosphatemia. $1,25(OH)_2D$ (calcitriol), the active vitamin D metabolite, is largely bound to vitamin D-binding protein in the serum (DBP). Free calcitriol enters the target cell, where it binds and activates the nuclear vitamin D receptor (VDR), stimulating the transcription of vitamin D-sensitive genes. The primary target tissues for vitamin D action are the small intestine, bone, and kidney.

# Hyperparathyroidism 1

(also known as: HRPT1; familial isolated primary hyperparathyroidism [FIHP]; familial parathyroid adenoma)

| | |
|---|---|
| **MIM** | 145000 |

**Clinical features**   Patients present with primary hyperparathyroidism causing hypercalcemia, hypophosphatemia, and elevated parathyroid hormone levels. Most have low or low to normal magnesium levels. Elevated magnesium levels suggest the presence of a calcium-sensing receptor (CaSR) mutation. Most patients have multiple gland involvement. On histologic examination, cystic parathyroid tumors are seen in 24% of patients. Parathyroid carcinoma, which is particularly common in patients with hyperparathyroidism–jaw tumor syndrome (HRPT–JT; MIM 145001, next entry), is rarely seen in HRPT1. A detailed clinical evaluation may reveal the presence of some nonparathyroid tumors, including breast cancer, colon cancer, and lipomas. However, no association has been reported that is consistent or distinctive enough to define a specific clinical category.

**Primary tissue or gland affected**   Parathyroid.

**Age of onset**   The average age of diagnosis of HRPT1 in probands is $39 \pm 3$ years. Family screening will result in earlier diagnosis.

**Epidemiology**   Familial hyperparathyroidism syndromes are rare, but should be sought in families with an appropriate family history and in patients with young-onset disease. The female to male ratio (1.5:1) is lower than that reported for sporadic hyperparathyroidism (2.7:1).

**Inheritance**   Autosomal dominant.

| | |
|---|---|
| **Chromosomal location** | 1q25-q31 (HRPT–JT), 11q13 (menin), 3q (CaSR), 19p (hypocalciuric hypercalcemia, familial benign, type 2 [HHC2]), 19q (HHC3). |
| **Genes** | In a recent review of 36 kindreds with apparent FIHP, *CASR* gene mutations were identified in five, and three had evidence of HRPT–JT syndrome. The remaining families were too small to allow linkage analysis. However, in previous publications of larger kindreds with HRPT1, the disease was successfully linked in all kindreds to one of the previously described syndromic loci: multiple endocrine neoplasia (MEN)1 (MIM 131100, Ch11q13); CaSR (MIM 601199, Ch3q); HRPT–JT (MIM 145001, Ch1q25-q31); HHC2 (MIM 145981, Ch19p); and HHC3 (MIM 600740, Ch19q). This suggests that some, if not all, families with HRPT1 in fact have one of these other syndromes. |
| **Mutational spectrum** | Unknown. |
| **Effect of mutation** | Unknown. |
| **Diagnosis** | Patients with primary parathyroid hyperplasia must be evaluated for the presence of other endocrine tumors, particularly those associated with the MEN1 and MEN2 syndromes. MEN2a families rarely present with familial isolated hyperparathyroidism, since medullary thyroid carcinoma and pheochromocytoma usually dominate the clinical picture. MEN1 families may present clinically with isolated hyperparathyroidism, but careful evaluation of all affected family members almost always reveals evidence of pituitary or pancreatic endocrine tumors. Careful radiologic evaluation of the jaw is necessary, and the finding of a jaw tumor (typically cemento-ossifying fibromas) strongly suggests HRPT–JT syndrome. The high incidence of parathyroid carcinoma in HRPT–JT syndrome makes this specific diagnosis clinically important. *CASR* mutations should be excluded, and affected family members should be evaluated for hypocalciuria, the mechanism of which is as yet unclear. Only when |

all of these evaluations are negative can the family be presumed to have HRPT1. However, asynchronous appearance of different aspects of MEN1 suggests that continued follow-up is needed. Genetic linkage analysis suggests that a significant percentage of families with HRPT1 may in fact have mutations in the *HRPT–JT* gene that result in partial penetrance.

**Counseling issues**

This is a treatable autosomal dominant disease. Because of the reports of parathyroid carcinoma, long-term follow-up is necessary for all patients.

**Notes**

As mentioned above, HRPT1 is likely to be a mixed group of genetic diseases, some or all of which may be variants of other hyperparathyroid syndromes. Careful clinical evaluation and follow-up may identify families with MEN1 or HRPT–JT syndrome. Once the precise genetic etiology of HRPT–JT syndrome is discovered, careful genetic evaluation of HRPT1 families is likely to prove fruitful. The possibility that mutations in genes at as yet unidentified loci can cause this syndrome cannot be excluded.

**References**

Kassem M, Zhang X, Brask S et al. Familial isolated primary hyperparathyroidism. *Clin Endocrinol (Oxf)* 1994;41:415–20.

Simonds WF, James-Newton LA, Agarwal SK et al. Familial isolated hyperparathyroidism: Clinical and genetic characteristics of 36 kindreds. *Medicine (Baltimore)* 2002;81:1–26 (Review).

Teh BT, Esapa CT, Houlston R et al. A family with isolated hyperparathyroidism segregating a missense MEN1 mutation and showing loss of the wild-type alleles in the parathyroid tumors. *Am J Hum Genet* 1998;63:1544–9 (Letter).

Teh BT, Farnebo F, Twigg S et al. Familial isolated hyperparathyroidism maps to the hyperparathyroidism–jaw tumor locus in 1q21-q32 in a subset of families. *J Clin Endocrinol Metab* 1998;83:2114–20.

# Hyperparathyroidism–Jaw Tumor Syndrome

(also known as: HRPT–JT; familial hyperparathyroidism 2 [HRPT2]; familial
hyperparathyroidism with ossifying jaw fibromas; familial cystic parathyroid adenomatosis)

| | |
|---|---|
| MIM | 145001 |
| Clinical features | Patients present with a clinical endocrine picture identical to that of sporadic hyperparathyroidism: hypercalcemia, hypophosphatemia, and hypercalciuria. Histologic evaluation may show parathyroid hyperplasia, and multiple cystic parathyroid adenomas are frequently seen. The adenomas may not appear at the same time, and removal of one may result in apparent cure, albeit temporary. The associated jaw tumors are typically fibrous maxillary or mandibular tumors resembling ossifying fibromas (cemento-ossifying fibromas), and not brown tumors as seen in primary hyperparathyroidism. Parathyroid carcinoma is common in this syndrome, and occult HRPT–JT may be responsible for most cases of parathyroid malignancy. Renal cysts are also commonly found, and there may be an increased incidence of Wilms' tumor. |
| Primary tissue or gland affected | Parathyroid. |
| Age of onset | Diagnosis is usually made during the second or third decade of life. |
| Epidemiology | This syndrome is rarely diagnosed, but it may be the main cause of parathyroid carcinoma and a major cause of 'isolated' familial hyperparathyroidism (HRPT1; MIM 145000). |
| Inheritance | Autosomal dominant. |
| Chromosomal location | 1q25-q31 |
| Genes | Unknown. |

| **Mutational spectrum** | Unknown. |
| **Effect of mutation** | Unknown. |

**Diagnosis**

The diagnosis is based on the finding of hyperparathyroidism due to parathyroid hyperplasia or multiple adenomas in a patient with typical jaw tumors. Jaw tumors may be occult, emphasizing the importance of routine X-ray evaluation of the jaw in patients with young-onset disease, multigland involvement, cystic parathyroid adenomas, and parathyroid carcinoma. All patients with young-onset hyperparathyroidism, multigland involvement, or positive family history should be carefully evaluated for all of the genetic syndromes associated with hyperparathyroidism, particularly multiple endocrine neoplasia (MEN)1, MEN2, HRPT1, and HRPT–JT.

**Counseling issues**

Because of the increased incidence of parathyroid carcinoma, long-term follow-up is essential.

**Notes**

The associated tumors frequently show loss of heterozygosity for the chromosome 1 locus, suggesting that the gene responsible is a tumor suppressor.

**References**

Fujikawa M, Okamura K, Sato K et al. Familial isolated hyperparathyroidism due to multiple adenomas associated with ossifying jaw fibroma and multiple uterine adenomyomatous polyps. *Eur J Endocrinol* 1998;138:557–61.

Szabo J, Heath B, Hill VM et al. Hereditary hyperparathyroidism–jaw tumor syndrome: The endocrine tumor gene HRPT2 maps to chromosome 1q21-q31. *Am J Hum Genet* 1995;56:944–50.

# Familial Hypocalciuric Hypercalcemia Type 1

(also known as: FHH1; hypocalciuric hypercalcemia type 1 [HHC1]; neonatal severe primary hyperparathyroidism [NSHPT])

| | |
|---|---|
| **MIM** | 145980 (HHC1), 239200 (NSHPT) |

**Clinical features**
Patients present with mild, asymptomatic hypercalcemia, which is usually found during evaluation for some unrelated reason. Additional clinical features include hypermagnesemia, normal or mildly elevated parathyroid hormone, and low urinary calcium excretion. The low urinary calcium excretion is caused by decreased calcium reabsorption in the ascending limb of Henle's loop. In this region of the nephron, calcium inhibits potassium channel activity. This action is mediated by the calcium-sensing receptor (CaSR) through an arachidonic acid second messenger. Decreased potassium channel activity alters the electrical gradient and thus decreases calcium reabsorption. Decreased function of CaSR will have the opposite effect, increasing calcium reabsorption and thus causing hypocalciuria, while increasing intravascular calcium. FHH1 is benign, in that calcium levels are only minimally elevated, and there are no apparent symptoms related to the disorder. However, a few patients have been reported with pancreatitis. The importance of identifying this disorder lies in differentiating this condition, which requires no therapy, from primary hyperparathyroidism, which frequently requires surgical intervention. Patients homozygous for these mutations present with severe neonatal hyperparathyroidism. Interestingly, this syndrome may be caused by a *de novo* mutation in one allele of the receptor that appears to function as a dominant negative, interfering with the function or expression of the normal allele.

**Primary tissue or gland affected**
Parathyroid.

| | |
|---|---|
| **Age of onset** | The defect is present from birth, but is benign and is diagnosed only when serum and/or urinary calcium levels are measured for some unrelated reason. |
| **Epidemiology** | Rare. |
| **Inheritance** | Autosomal dominant. |
| **Chromosomal location** | 3q13.3-q21 |
| **Genes** | *CASR* (calcium-sensing receptor). |
| **Mutational spectrum** | Insertion, deletion, and missense mutations have been described and are located throughout the molecule. Most are on the N-terminal, extracellular tail. |
| **Effect of mutation** | These are inactivating mutations, resulting in a decreased affinity for calcium, and thus decreased signal transduction at any given calcium level. In some cases, the mutation may result in fewer normally functioning receptors on the surface of parathyroid and renal cell membranes. |
| **Diagnosis** | The clinical diagnosis is made by finding inappropriately low urinary calcium excretion (calcium clearance to creatinine clearance ratio <0.01) in the presence of hypercalcemia with normal or minimally elevated parathyroid levels. It may often be difficult to differentiate this from mild hyperparathyroidism associated with a very low calcium intake or concomitant vitamin D deficiency. A positive family history consistent with autosomal dominant inheritance may be particularly helpful. Definitive diagnosis requires the identification of a mutation in the *CASR* gene. These mutations can be anywhere in the gene, and most families have their own, individual mutation, making molecular diagnosis difficult. |

**Counseling issues**     This is a benign disorder. Definitive diagnosis is important to avoid unnecessary surgery. For this reason, family screening is recommended.

**References**     Chou YH, Brown EM, Levi T et al. The gene responsible for familial hypocalciuric hypercalcemia maps to chromosome 3q in four unrelated families. *Nat Genet* 1992;1:295–300.

Thakker RV. Disorders of the calcium-sensing receptor. *Biochim Biophys Acta* 1998;1448:166–70.

# Familial Hypocalciuric Hypercalcemia Types 2 and 3

(also known as: FHH2 and FHH3; hypocalciuric hypercalcemia, familial types 2 and 3 [HHC2 and HHC3]; familial benign hypercalcemia types 2 and 3 [FBH2 and FBH3]; familial benign hypercalcemia, Oklahoma variant [FBHOk])

| | |
|---|---|
| **MIM** | 145981 (HHC2), 600740 (HHC3) |
| **Clinical features** | The proband of a pedigree is usually identified when hypercalcemia is found during a routine examination, or during evaluation for an unrelated problem. Urinary calcium excretion, performed as part of the evaluation of hyperparathyroidism, is inappropriately low, suggesting the diagnosis of HHC. In HHC2, parathyroid hormone (PTH) levels typically remain normal. In contrast, in HHC3, PTH levels are usually normal in adolescents and young adults, but appear to increase with age, and may become elevated during the fourth decade of life. Hypophosphatemia and osteomalacia may also appear in individuals >40 years old. |
| **Primary tissue or gland affected** | Parathyroid. |
| **Age of onset** | Hypercalcemia may be found in childhood, and is usually discovered as an incidental finding during routine evaluation for an unrelated complaint, or during family screening. |
| **Epidemiology** | Several kindreds have been described with biochemical evidence of HHC in which linkage to the Ch3q21-24 locus has been excluded. In some, linkage to 19p13.3 can be demonstrated. In a large kindred from Oklahoma, both loci were excluded by linkage, and a third locus, 19q13, was identified as containing the causative gene. |
| **Inheritance** | Autosomal dominant. |
| **Chromosomal location** | 19p13.3 (HHC2), 19q13 (HHC3). |

| | |
|---|---|
| **Genes** | Unknown. |
| **Mutational spectrum** | Unknown. |
| **Effect of mutation** | Because of the clinical findings, it is suspected that the mutant gene is either a cofactor of the calcium-sensing receptor (CaSR), or a downstream mediator in the CaSR pathway. Alternatively, it may be part of an independent, parallel-acting calcium-sensing mechanism. |
| **Diagnosis** | The clinical diagnosis is made when hypercalcemia is found in the presence of relative hypocalciuria. A calcium to creatinine clearance of <0.01 is typical. HHC is a genetically heterogeneous entity. Family studies are consistent with autosomal dominant inheritance. The diseases caused by each of the three loci identified to date are phenotypically very similar, and cannot be differentiated clinically. The elevated PTH levels found in older patients with HHC3 make differentiation from primary hyperparathyroidism particularly difficult. Mutation analysis of the *CASR* gene, or linkage analysis if the pedigree is sufficiently large, is necessary for complete genetic classification. |
| **Counseling issues** | Parathyroid surgery, the preferred treatment for hyperparathyroidism, is not indicated in these patients. Therefore, accurate diagnosis may prevent unnecessary surgery in affected individuals. |
| **References** | Heath HD 3rd, Leppert MF, Lifton RP et al. Genetic linkage analysis in familial benign hypercalcemia using a candidate gene strategy. I. Studies in four families. *J Clin Endocrinol Metab* 1992;75:846–51. |
| | Lloyd SE, Pannett AA, Dixon PH et al. Localization of familial benign hypercalcemia, Oklahoma variant (FBHOk), to chromosome 19q13. *Am J Hum Genet* 1999;64:189–95. |

# Familial Hypercalciuric Hypercalcemia

**MIM**   601199

**Clinical features**   The clinical syndrome consists of hypercalcemia, hypercalciuria, hypermagnesemia, and inappropriately high parathyroid hormone levels. In the one family described so far, most patients had chief-cell parathyroid hyperplasia, whereas one was found to have a single parathyroid adenoma. Two patients had a history of renal stones. This syndrome is distinct from familial hypocalciuric hypercalcemia type 1 (FHH1; MIM 145980, p. 127), which is not associated with renal or other hypercalcemic symptoms. Differentiation between this syndrome and familial hyperparathyroidism type 1 (MIM 145000, p. 122) may be particularly difficult because hypercalciuria is seen in both. The presence of hypermagnesemia in a patient with primary hyperparathyroidism suggests the presence of a calcium-sensing receptor (CaSR) mutation.

**Primary tissue or gland affected**   Parathyroid.

**Age of onset**   In the family described, diagnosis was made when the subjects were between 23 years and 65 years of age. Family members <18 years old were not tested. Therefore, it may be possible to confirm the biochemical diagnosis at an earlier age.

**Epidemiology**   A single large family has been reported with this disorder.

**Inheritance**   Dominant.

**Chromosomal location**   3q13.3-q21

**Genes**   *CASR* (calcium-sensing receptor).

**Mutational spectrum**     A point mutation (F881L) has been found in the cytoplasmic tail of the *CASR* gene. The location of this mutation is different from those associated with FHH1 and autosomal dominant familial hypocalcemia (MIM 146200), where mutations are located in the extracellular and transmembrane portions of the molecule.

**Effect of mutation**      Inactivating mutation of CaSR.

**Diagnosis**               This disease must be suspected if there is clinical evidence of familial hyperparathyroidism without associated signs or symptoms that suggest the diagnosis of multiple endocrine neoplasia (MEN)1 (MIM 131100), MEN2 (MIM 171400), or of hyperparathyroidism–jaw tumor syndrome (MIM 145001).

**Notes**                   At least one patient in the family was reported to have a parathyroid adenoma. Patients respond well to 'radical' parathyroidectomy.

**References**              Carling T, Szabo E, Bai M et al. Familial hypercalcemic hypercalciuria caused by a novel mutation in the cytoplasmic tail of the calcium receptor. *J Clin Endocrinol Metab* 2000;85:2042–7.

# Metaphyseal Chondrodysplasia, Jansen Type

(also known as: parathyroid hormone receptor-activating mutation)

| | |
|---|---|
| **MIM** | 156400 |

**Clinical features**
Patients present with severe short stature and severe skeletal deformities due to metaphyseal dysplasia. Petrous bone sclerosis may lead to deafness. Hypercalcemia, hypercalciuria, and hypophosphatemia are noted in childhood, though the degree of hypophosphatemia is variable. Parathyroid hormone (PTH) and PTH-related protein (PTH-RP) levels are suppressed, while urinary cAMP and 1,25-dihydroxyvitamin D levels are elevated. Biochemical indices of bone metabolism reveal increased resorption without compensatory increased formation.

**Primary tissue or gland affected**
Parathyroid.

**Other organs, tissues, or glands affected**
Bone.

**Age of onset**
Patients are born with intrauterine growth retardation, presumably caused by constitutive activation of the PTH/PTH-RP receptor during fetal development. Dysmorphic features due to metaphyseal dysplasia may be present at birth, or may become evident during the first years of life.

**Epidemiology**
The syndrome is extremely rare, with only a very few families reported in the literature. Some cases are caused by *de novo* mutations.

**Inheritance**
Autosomal dominant.

**Chromosomal location**
3p22-p21.1

| Genes | PTHR (parathyroid hormone receptor). |
|---|---|

| Mutational spectrum | Specific point mutations at codons 223 and 410 have been shown to produce constitutively active PTH receptors. *In vitro* studies show that substitution of histidine with either arginine or lysine at position 223 leads to constitutive receptor activity. Other amino acid substitutions either modify surface expression of the receptor or have no effect on function. In contrast, most amino acid substitutions at position 410 lead to constitutive receptor activity. |
|---|---|

| Effect of mutation | The mutations result in the expression of constitutively active receptors in PTH and PTH-RP target tissues. |
|---|---|

| Diagnosis | Biochemical evaluation reveals changes consistent with severe hyperparathyroidism, but with an absence of circulating PTH. Similarly, PTH-RP is absent in the serum. This clinical picture is consistent with either the presence of a circulating factor that activates PTH receptors in an unregulated manner, or activating mutations in the receptor itself. Definitive diagnosis is made by molecular analysis of the *PTHR* gene. |
|---|---|

| Counseling issues | Both male and female patients are fertile. Once the mutation is identified, prenatal diagnosis can be offered. |
|---|---|

| References | Calvi LM, Schipani E. The PTH/PTHrP receptor in Jansen's metaphyseal chondrodysplasia. *J Endocrinol Invest* 2000;23:545–54 (Review). |
|---|---|
| | Campbell JB, Kozlowski K, Lejman T et al. Jansen type of spondylometaphyseal dysplasia. *Skeletal Radiol* 2000;29:239–42. |
| | Schipani E, Jensen GS, Pincus J et al. Constitutive activation of the cyclic adenosine 3´,5´-monophosphate signaling pathway by parathyroid hormone (PTH)/PTH-related peptide receptors mutated at the two loci for Jansen's metaphyseal chondrodysplasia. *Mol Endocrinol* 1997;11:851–8. |

# Williams' Syndrome

(also known as: WS; Williams–Beuren syndrome; 'Elfin' facies; hypercalcemia in infancy; supravalvular aortic stenosis)

| | |
|---|---|
| MIM | 194050 |

**Clinical features**　Children with WS have a characteristic 'elfin' facial appearance: flat nasal bridge, anteverted nares, puffy eyes, epicanthal folds, a small mandible, and prominent cheeks. Premature graying of the hair occurs in young adults. Postnatal growth retardation and early onset of puberty result in a short final height (mean heights: men, 165 cm; women, 152 cm). The most common cardiovascular manifestations are supravalvular aortic stenosis and peripheral pulmonary stenosis. Hypercalcemia in infancy may contribute to irritability, vomiting, constipation, and muscle cramps. Hypercalcemia is usually transient, but hypercalciuria may persist and lead to nephrocalcinosis. Cognitive, motor, and language delays are noted in all WS patients; mild to severe mental retardation is present in about 75%. Personality characteristics are described as overly friendly, but anxious, and lacking social judgment. Other manifestations include renal anomalies, strabismus and esotropia, and orthopedic problems (kyphosis, scoliosis, and lordosis). Hypersensitivity to certain sounds (hyperacusis) is present in about 90% of WS individuals. Constipation and rectal prolapse occur in infancy; diverticulitis and peptic ulcer are noted in older patients.

**Primary tissue or gland affected**　Parathyroid.

**Age of onset**　Onset is at birth, but children are usually diagnosed in infancy or childhood.

**Epidemiology**　The incidence of WS is estimated to be 1:20 000 live births. Both sexes are equally affected.

| | |
|---|---|
| **Inheritance** | Most cases of WS are sporadic, but some familial cases with an autosomal dominant pattern have been observed. |
| **Chromosomal location** | 7q11.23 |
| **Genes** | *ELN* (elastin), *LIMK1* (LIM kinase 1). Other genes in the WS region of 7q11.23 have been described, but their role in producing the clinical syndrome is less clear. |
| **Mutational spectrum** | A contiguous gene deletion at 7q11.23 is present in 99% of WS patients. Approximately 17 genes are included in the 1.5 Mb deletion. |
| **Effect of mutation** | *ELN* mutations lead to structurally abnormal tropoelastin, which may affect the elasticity of skin, lungs, and large blood vessels. Elastin is the major extracellular matrix protein and comprises half of the dry weight of the aorta. Hemizygosity for *ELN* correlates well with supravalvular aortic stenosis and other vascular stenoses, but does not explain the other features of WS. *LIMK1* encodes a tyrosine kinase that inactivates cofilin, which is critical for the recycling of actin filaments. *LIMK1* mutations may result in the formation of abnormal axonal pathways during central nervous system development. |
| **Diagnosis** | Patients who display the clinical features of WS should undergo chromosome analysis and fluorescence *in situ* hybridization to confirm the diagnosis. |
| **Counseling issues** | Because the variable manifestations of WS appear and/or resolve at different stages of life, screening tests for vision, cardiac, orthopedic, and gastrointestinal manifestions should be repeated at regular intervals from infancy through adulthood. |

**References**

American Academy of Pediatrics. Health care supervision for children with William's syndrome. *Pediatrics* 2001;107:1192–204.

Donnai D, Karmiloff-Smith A. William's syndrome: From genotype through to the cognitive phenotype. *Am J Med Genet* 2000;97:164–71 (Review).

# Pseudohypoparathyroidism Type 1a

(also known as: PHP1a; Albright hereditary osteodystrophy [AHO]; pseudopseudohypoparathyroidism)

| | |
|---|---|
| **MIM** | 103580 |
| **Clinical features** | Patients present with hypocalcemia and hyperphosphatemia suggestive of hypoparathyroidism. However, unlike true hypoparathyroidism, parathyroid hormone (PTH) levels are elevated, and urinary cAMP response to exogenous PTH is reduced. The latter suggests the presence of a defect in responsiveness to PTH, and not secretion of inactive hormone. The typical phenotype of PHP1a includes short stature, obesity, round facies, low nasal bridge, and short metacarpals and metatarsals (especially the fourth and fifth). Some degree of mental retardation is present in the majority of patients. Some patients may present with other hormone deficiencies, such as hypothyroidism and hypogonadism, caused by resistance to other hormones that act through cAMP-dependent pathways. |
| **Primary tissue or gland affected** | Parathyroid. |
| **Other organs, tissues, or glands affected** | Thyroid, gonads. |
| **Age of onset** | The defect is present at birth, though most patients are diagnosed during mid-childhood. |
| **Epidemiology** | PHP1a is rare; the precise incidence has not yet been determined. |
| **Inheritance** | Autosomal dominant (imprinted). |
| **Chromosomal location** | 20q13.2 |

| Genes | GNAS1 (stimulatory Gs-$\alpha$-1 subunit of adenylyl cyclase). |
|---|---|
| Mutational spectrum | More than 26 different missense, termination, and deletion mutations have been described. |
| Effect of mutation | PHP1a is caused by loss of function mutations. The gene is imprinted in a tissue-specific manner, and renal expression is limited to the maternal allele. Mutations in the GNAS1 gene have been identified. Interestingly, this gene is differently imprinted in different tissues. Most tissues express both alleles and, thus, mutations in either allele will cause haploinsufficiency, resulting in the typical physical characteristics. However, in the kidney, the maternally inherited allele is preferentially expressed. Therefore, a patient who inherits a mutant allele from his mother will have the skeletal as well as the renal characteristics of the disease. If, however, the mutation is inherited from the father, the skeletal defects will be present, but renal response to PTH will be normal. This syndrome is known as pseudo-PHP since the patient appears to have all of the phenotypic characteristics of PHP1a, but normal calcium, phosphorus, PTH, and urinary cAMP levels. Thus, in the same pedigree, in one branch the mutation may be inherited from the mother, and the offspring have PHP1a, whereas in another branch of the family the mutation may be inherited from a male carrier, and the offspring have pseudo-PHP. |
| Diagnosis | The diagnosis is suspected clinically when a patient presents with hypocalcemia and the typical skeletal abnormalities. It can be confirmed biochemically by demonstrating hypocalcemia, hyperphosphatemia, elevated PTH, and a low urinary cAMP response to exogenous PTH. Mutation analysis of the GNAS1 gene will confirm the diagnosis and permit genetic counseling. |
| Counseling issues | In most families, the disease is inherited as an autosomal dominant trait as described above. Some cases of de novo mutations have been described. |

**Notes**

Note that the phenotypic characteristics of PHP can be divided into two groups. The first is the physical phenotype, including short stature, obesity, round facies, and short fourth and fifth metacarpals and metatarsals. The second is the renal–electrolyte phenotype, including hypocalcemia, hyperphosphatemia, elevated PTH, and a decreased urinary cAMP response to exogenous PTH. The complete syndrome (PHP1a) is caused by maternally inherited *GNAS1* mutations. Paternal inheritance of the same mutation causes pseudo-PHP1, in which the physical characteristics of PHP1 are present, but PTH function is normal. Other mutations in the same chromosomal locus, perhaps promoter or imprint mutations of the *GNAS1* gene, cause the renal–electrolyte abnormalities without the physical phenotype.

PHP1b is biochemically different, but may be caused by mutations in the same gene (see the following entry).

# Pseudohypoparathyroidism Type 1b

(also known as: PHP1b)

| | |
|---|---|
| **MIM** | 603233 |

**Clinical features**  Patients with this variant of PHP type 1 lack the physical features of PHP1a, but show the typical signs of renal resistance to parathyroid hormone (PTH), including hypocalcemia, hyperphosphatemia, elevated PTH levels, and lack of cAMP response to exogenous PTH. Some patients may manifest skeletal lesions suggestive of hyperparathyroidism. Taken together, these findings suggest an organ-specific resistance to PTH limited to the kidney.

**Primary tissue or gland affected**  Parathyroid.

**Age of onset**  The initial clinical symptoms of hypocalcemia typically present at approximately 8 years of age.

**Epidemiology**  PHP1b is very rare, comprising a small subset of pseudohypoparathyroidism, which is in itself quite rare.

**Inheritance**  Autosomal dominant.

**Chromosomal location**  20q13.3

**Genes**  GNAS1 (stimulatory Gs-α-1 subunit of adenylyl cyclase) mutations have not been found, and Gs-α activity is normal in the tissues tested. Genetic analysis has established linkage to Ch20q13.3, the region containing the GNAS1 gene, though the highest linkage score is located slightly centromeric to the GNAS1 gene. It is possible that the mutation involves a disruption of the promoter or of the imprinting region, resulting in a specific renal defect. Alternatively, the disease may be caused by another gene located in this region.

| | |
|---|---|
| **Mutational spectrum** | Unknown. |
| **Effect of mutation** | Unknown. |
| **Diagnosis** | The diagnosis is based on demonstrating biochemical characteristics typical of PHP1 in the absence of the physical characteristics of PHP1a. Normal Gs activity confirms PHP1b and excludes PHP1a. |
| **Counseling issues** | This is a treatable disease, and early diagnosis and treatment are likely to improve the long-term prognosis. |
| **Notes** | The phenotypic characteristics of PHP can be divided into two groups; see Notes in the previous entry. |

# Pseudohypoparathyroidism Type 2

(also known as: PHP2)

| | |
|---|---|
| MIM | 203330 |
| Clinical features | Patients present with resistance to parathyroid hormone (PTH) characterized by hypocalcemia, hyperphosphatemia, and elevated PTH levels. Physical appearance is normal, and there is no involvement of any other endocrine glands. Correction of the hypocalcemia by calcium and vitamin D supplementation can also correct phosphaturia in some cases. |
| Primary tissue or gland affected | Parathyroid. |
| Age of onset | The initial patients described with this disorder presented at the age of 22 months with seizures, suggesting that this form of PHP may be more severe, and may present earlier than the other forms described above. Too few cases have been reported to provide an accurate description of the phenotypic variability. |
| Epidemiology | Only 25 cases have been reported in the literature. Most are sporadic, though at least one case of familial disease has been reported. |
| Inheritance | Unknown. |
| Chromosomal location | Unknown. |
| Genes | Unknown. |
| Mutational spectrum | Unknown. |

| | |
|---|---|
| **Effect of mutation** | PTH acts by binding to its receptor and activating adenylyl cyclase, which causes an increase in intracellular cAMP. This in turn activates downstream effector pathways that are incompletely understood. The clinical finding of a normal cAMP, but abnormal phosphaturic, response to exogenous PTH suggests that the defect must be in the signal transduction pathways distal to cAMP. |
| **Diagnosis** | PHP is diagnosed by finding electrolyte abnormalities typical of hypoparathyroidism (hypocalcemia and hyperphosphatemia) in the presence of elevated PTH levels. PHP2 can be differentiated from PHP1 by measuring the cAMP response to exogenous PTH. In the latter, the cAMP response to exogenous PTH is absent or greatly reduced, whereas in PHP2 there is an apparently normal response. These findings suggest that the metabolic defect involves a second messenger system distal to, or independent of, cAMP. |
| **Counseling issues** | Too little is known about the genetics of this disorder to allow accurate prediction of genetic risk. |
| **Notes** | Profound hypocalcemia due to severe vitamin D deficiency may cause a similar biochemical syndrome, with elevated PTH and nephrogenous cAMP, but abnormal tubular reabsorption of phosphate. Thus, some patients with apparent PHP2 may in fact have severe vitamin D deficiency. |
| **References** | Levine MA. Clinical spectrum and pathogenesis of pseudohypoparathyroidism. *Rev Endocr Metab Disord* 2000;1:265–74 (Review). |

# Familial Isolated Hypoparathyroidism

(also known as: FIH)

| | |
|---|---|
| **MIM** | 146200 |

**Clinical features**

Serum calcium levels are low, typically in the 6–8 mg/dL (1.5–2 mmol/L) range. Urinary calcium secretion is inappropriately high, and serum parathyroid hormone (PTH) levels are usually normal. Pearce et al. reported that the mean (±SE) ratio of urinary calcium to urinary creatinine (expressed as milligrams of calcium per milligram of creatinine) was 0.16 ± 0.02 (0.5 ± 0.1 mmol Ca per mmol creatinine) as compared with 0.07 ± 0.02 (0.2 ± 0.1 mmol Ca per mmol creatinine) in hypoparathyroid patients with similar calcium levels. Patients are typically asymptomatic. A family history of hypocalcemia and recurrent nephrolithiasis is highly suggestive of this disorder. Some patients have symptomatic hypocalcemia, and some have low serum PTH concentrations. Magnesium levels may be low. Since most patients are asymptomatic, often no treatment is needed. When treatment is required, it is aimed at raising calcium levels to alleviate the symptoms. Treatment should be carefully monitored, since calcium and vitamin D supplementation are associated with marked hypercalciuria, leading to nephrocalcinosis, nephrolithiasis, and renal insufficiency.

**Primary tissue or gland affected**

Parathyroid.

**Age of onset**

Most patients are asymptomatic, so diagnosis will usually be made when calcium levels are measured for an unrelated reason.

**Epidemiology**

Rare.

**Inheritance**

Autosomal dominant.

**Chromosomal location**

3q13

| | |
|---|---|
| **Genes** | *CASR* (calcium-sensing receptor). |
| **Mutational spectrum** | Missense mutations have been described. Most are located in the extracellular N-terminal portion of the molecule. Two mutations have been identified in the transmembrane region. |
| **Effect of mutation** | These are activating mutations of the calcium-sensing receptor. The receptor negatively regulates calcium resorption by kidney cells. Activation of the receptor results in decreased resorption and hypercalciuria, even in the presence of hypocalcemia. At the level of the parathyroid, receptor activation inhibits the secretion of PTH. |
| **Diagnosis** | The clinical diagnosis is made by finding low serum calcium in the face of normal or elevated urinary calcium excretion. A major clinical indicator is the familial nature of the disease. As opposed to other causes of hypocalcemia, PTH levels are typically normal, though some cases with low PTH levels have been reported. Identification of an appropriate mutation in the *CASR* gene confirms the diagnosis. |
| **Counseling issues** | Patients with this disorder should only rarely be treated because of the danger of renal damage secondary to hypercalciuria, and any treatment should be monitored with particular care. Therefore, precise diagnosis is critical in order to avoid an attempt at normalizing calcium levels, as is typically done in patients with primary hypoparathyroidism. It should be noted that even in patients with primary hypoparathyroidism, hypercalciuria should be monitored and calcium levels should be maintained at the lowest level possible to alleviate symptoms. |
| **Notes** | It is important to emphasize the potential danger associated with vitamin D treatment and the resultant hypercalciuria. In all patients with hypocalcemia, treatment should be aimed at raising calcium levels to the point where the patient is asymptomatic, without causing hypercalciuria. Thiazide diuretics, which decrease calcium excretion, have been successfully used in patients with primary |

hypoparathyroidism. Their efficacy in patients with activating *CASR* mutations has not been documented.

**References**

Pollak MR, Brown EM, Estep HL et al. Autosomal dominant hypocalcaemia caused by a Ca(2+)-sensing receptor gene mutation. *Nat Genet* 1994;8:303–7.

Thakker RV. Disorders of the calcium-sensing receptor. *Biochim Biophys Acta* 1998;1448:166–70.

Pearce SH, Williamson C, Kifor O et al. A familial syndrome of hypocalcemia with hypercalciuria due to mutations in the calcium-sensing receptor. *N Engl J Med* 1996;335:1115–22.

# Hypoparathyroidism, X-linked

(also known as: HYPX; agenesis of parathyroid glands)

| | |
|---|---|
| MIM | 307700 |
| Clinical features | The clinical presentation is similar to any form of hypoparathyroidism, with severe hypocalcemia presenting as neonatal tetany. Parathyroid hormone (PTH) levels are unmeasurably low. An extensive autopsy performed on one affected individual failed to reveal any evidence of parathyroid tissue. |
| Primary tissue or gland affected | Parathyroid. |
| Age of onset | Birth. |
| Epidemiology | Extremely rare. Two Missouri families were initially reported. Although they were not known to be related, subsequent mitochandrial RNA analysis documented common ancestry. A single Bedouin family was also reported with multiple affected individuals. |
| Inheritance | X-linked, autosomal dominant. |
| Chromosomal location | Xq26-q27 |
| Genes | Unknown. |
| Mutational spectrum | Unknown. |
| Effect of mutation | Unknown. |
| Diagnosis | Diagnosis is based on signs of hypoparathyroidism in the neonate, including hypocalcemia, and hyperphosphatemia with no detectable PTH in the blood. Careful examination of the neck should fail to |

demonstrate any parathyroid tissue. The initial presentation may be similar to maternal hyperparathyroidism with fetal parathyroid suppression. This can be excluded by examining the mother, and by documenting persistence of hypoparathyroidism beyond the neonatal period.

**Counseling issues**

This is a treatable disease that can be fatal if not adequately addressed immediately after birth.

**References**

Mumm S, Whyte MP, Thakker RV et al. mtDNA analysis shows common ancestry in two kindreds with X-linked recessive hypoparathyroidism and reveals a heteroplasmic silent mutation. *Am J Hum Genet* 1997;60:153–9.

Trump D, Dixon PH, Mumm S et al. Localisation of X linked recessive idiopathic hypoparathyroidism to a 1.5 Mb region on Xq26-q27. *J Med Genet* 1998;35:905–9.

# Hypophosphatemic Rickets

(includes: X-linked hypophosphatemic rickets [X-HypR]; vitamin D-resistant
hypophosphatemic rickets; autosomal dominant hypophosphatemic rickets [ADHR];
Dent's disease [X-linked recessive nephrolithiasis, type 2])

| | |
|---|---|
| MIM | 307800, 193100, 300009 |
| Clinical features | Skeletal manifestations of X-HypR include ankylosis of the spine and major joints, spinal stenosis, bowed legs, and short stature. Nephrocalcinosis and sensorineural hearing loss may be present. Hypertension and left ventricular hypertrophy may result from the disease itself, or be secondary to treatment. Laboratory features include hypophosphatemia and elevated alkaline phosphatase, but serum calcium levels are normal. Although hypophosphatemia stimulates production of 1,25-dihydroxyvitamin D, levels of calcitriol are not elevated in X-HypR; this impairment of calcitriol production appears to be secondary to abnormal phosphate transport. Patients with ADHR may present in childhood with deformities of the lower extremities and other features of rickets, or after puberty with bone pain and fractures. The clinical picture of Dent's disease is variable, including mild to severe rickets, nephrolithiasis, and end-stage renal disease. Hypercalciuria, along with renal glycosuria, aminoaciduria, and phosphate wasting, may be variable and intermittent. |
| Primary tissue or gland affected | Bone. |
| Other organs, tissues, or glands affected | Kidney. |
| Age of onset | Childhood or adulthood. |
| Epidemiology | In X-HypR, males and females are affected, consistent with dominant X-linked inheritance. Fibroblast growth factor 23 |

mutations causing ADHR are rare; three mutations have been described in four families.

| | |
|---|---|
| **Inheritance** | X-linked dominant, autosomal dominant, X-linked recessive. |
| **Chromosomal location** | Xp22.2-p22.1, 12p13.3, Xp11.22. |
| **Genes** | *PHEX* (phosphate-regulating gene homologous to endopeptidases) mutations lead to X-HypR; *FGF23* (fibroblast growth factor 23) mutations are responsible for ADHR; *CLC5* (voltage-dependent chloride channel) mutations result in Dent's disease. |
| **Mutational spectrum** | Deletions, insertions, missense, nonsense, and splice-site mutations of *PHEX* have been described in families with X-HypR. Missense mutations affecting two amino acids of FGF23 have been reported in ADHR. Microdeletions in the *CLC5* gene have been identified in families with Dent's disease. |
| **Effect of mutation** | The precise mechanisms by which inactivating mutations of *PHEX* result in the clinical syndrome of X-HypR have not yet been documented. Several investigators have proposed that *PHEX* metabolizes 'phosphatonin', a putative circulating factor that regulates renal phosphate reabsorption and renal production of calcitriol. Mutant PHEX is less efficient than wild-type PHEX in degrading phosphatonin. The resulting increase in phosphatonin levels causes excessive renal tubular phosphate loss. Increased activity of *FGF23* in ADHR results from gain of function mutations, and may be mediated by impaired proteolytic cleavage. *FGF23* has been proposed as a possible candidate for 'phosphatonin'. *CLC5* is expressed in the proximal tubule of the kidney, in the thick ascending limb of Henle's loop, and in the acid-secreting α-intercalated cells of the collecting duct. Mutations in *CLC5* result in abnormal uptake and reabsorption of proteins and other solutes in the proximal tubule. |

| | |
|---|---|
| **Diagnosis** | The presence of rickets or osteomalacia in a child or adult with hypophosphatemia, elevated alkaline phosphatase, and normocalcemia suggests a diagnosis of hypophosphatemic rickets. The possibility of tumor-associated oncogenic osteomalacia needs to be excluded. |
| **Notes** | Treatment of X-HypR with phosphate alone is ineffective because phosphate lowers the plasma calcium concentration, resulting in a further decrease in calcitriol concentration, and causing secondary hyperparathyroidism that leads to worsening of the bone disease. Combined treatment with supplemental phosphate and calcitriol is effective in reversing rachitic changes and improving growth in prepubertal children. Calcitriol should not be used in those cases of hypophosphatemic rickets in which hypercalciuria is present, due to the increased risk of nephrocalcinosis. Abundant secretion of FGF23 by tumors has been implicated as a cause of oncogenic osteomalacia. |
| **References** | Quarles LD, Drezner MK. Pathophysiology of X-linked hypophosphatemia, tumor-induced osteomalacia, and autosomal dominant hypophosphatemia: A perPHEXing problem. *J Clin Endocrinol Metab* 2001;86:494–6. |
| | Scheinman SJ, Guay-Woodford LM, Thakker RV et al. Genetic disorders of renal electrolyte transport. *N Engl J Med* 1999;340:1177–87. |
| | White KE, Jonsson KB, Carn G et al. The autosomal dominant hypophosphatemic rickets (ADHR) gene is a secreted polypeptide overexpressed by tumors that cause phosphate wasting. *J Clin Endocrinol Metab* 2001;86:497–500. |

# Vitamin D-resistant Rickets

(also known as: vitamin D-dependent rickets [VDDR] types 2a and 2b;
rickets–alopecia syndrome; hypocalcemic vitamin D-resistant rickets)

| | |
|---|---|
| MIM | 277440, 277420 |
| Clinical features | In addition to rickets and osteomalacia, clinical manifestations include tetany and seizures due to hypocalcemia in infants and children, enamel hypoplasia and oligodontia, and, in some families, alopecia totalis. Hyperparathyroidism that is secondary to hypocalcemia results in hypophosphatemia and aminoaciduria. Levels of 25-hydroxyvitamin D levels are normal, but 1,25-dihydroxyvitamin D is markedly increased. Intestinal resistance to vitamin D causes impaired intestinal calcium absorption and high fecal calcium loss. Treatment includes pharmacologic doses of vitamin D metabolites and/or intravenous calcium infusions (to circumvent the impaired intestinal absorption), but responses vary. |
| Primary tissue or gland affected | Bone. |
| Other organs, tissues, or glands affected | The renal tubules, intestines, hair follicles (in VDDR type 2a), and teeth may be affected. |
| Age of onset | Infancy or childhood. |
| Epidemiology | This is a very rare condition. Patients have been described from Haitian, Arab, and Turkish backgrounds. |
| Inheritance | Autosomal recessive. |
| Chromosomal location | 12q12-q14 |
| Genes | VDR (vitamin D receptor). |

**Mutational spectrum**   More than 22 functional mutations have been reported, mostly located in the DNA-binding domain (DBD). Four mutations in the ligand-binding domain (LBD) have been described, of which two are stop mutations and one a missense mutation.

**Effect of mutation**   Mutations in the DBD have no effect on the ligand-binding properties of the receptor, and result in the receptor-positive phenotype. *In vitro* studies showed that the mutant receptors are unable to activate transcription of a reporter gene, even in the presence of very high concentrations of hormone. Stop mutations lead to the deletion of portions of the LBD. Other mutations have been shown to decrease receptor affinity by as much as 1000-fold.

**Diagnosis**   Elevated serum levels of 1,25-dihydroxyvitamin D in a patient with clinical and biochemical manifestations of rickets confirms the diagnosis of vitamin D resistance. Definitive diagnosis is based on genetic evaluation of the *VDR* gene.

**Notes**   Some patients have been observed to undergo clinical remission and maintain normal or near-normal biochemical parameters off therapy. The mechanism for this phenomenon is unexplained. Relapses may occur during pregnancy or intercurrent intestinal diseases.

**References**   Malloy PJ, Eccleshall TR, Gross C et al. Hereditary vitamin D resistant rickets caused by a novel mutation in the vitamin D receptor that results in decreased affinity for hormone and cellular hyporesponsiveness. *J Clin Invest* 1997;99:297–304.

Thomas MK, Demay MB. Vitamin D deficiency and disorders of vitamin D metabolism. *Endocrinol Metab Clin North Am* 2000;29:611–27, viii (Review).

# Vitamin D-dependent Rickets

**(also known as: VDDR type 1; 1α-hydroxylase deficiency; pseudovitamin D deficiency)**

| | |
|---|---|
| **MIM** | 264700 |
| **Clinical features** | Clinical manifestations of the failure to convert vitamin D to calcitriol include hypocalcemic seizures in infancy, and growth retardation, bowing of the legs, enamel hypoplasia of the teeth, and bone pain in early childhood. Laboratory findings include hypocalcemia, hypophosphatemia, and elevated serum alkaline phosphatase and parathyroid hormone (PTH) levels. Levels of 25-hydroxyvitamin D are normal; 1,25-dihydroxyvitamin D levels are low. |
| **Primary tissue or gland affected** | Bone. |
| **Other organs, tissues, or glands affected** | Intestinal absorption of calcium. |
| **Age of onset** | Clinical signs of VDDR usually appear within the first few months of life, and almost always before the age of 2 years. |
| **Epidemiology** | This disorder is common among French Canadians and extremely rare in other populations. In the Saguenay region of Quebec, the estimated prevalence is 1:2358 live births. |
| **Inheritance** | Autosomal recessive. |
| **Chromosomal location** | 12q14 |
| **Genes** | *P450c1*-α (1α-hydroxylase). |
| **Mutational spectrum** | Deletions, duplications, and missense mutations have been reported. |

**Effect of mutation**      All of the mutations reported to date result in complete absence of 1α-hydroxylase activity.

**Diagnosis**      The biochemical diagnosis is based on finding hypocalcemia in the presence of normal 25-hydroxyvitamin D levels, markedly reduced 1,25-dihydroxyvitamin D levels, and elevated PTH levels. Serum levels of 1,25-dihydroxyvitamin D are normal or elevated in rickets due to nutritional or genetic disorders. In addition to hereditary 1α-hydroxylase deficiency, low levels of this metabolite are found in renal osteodystrophy and hypoparathyroidism.

**Notes**      Patients show a rapid and sustained response to physiologic doses of calcitriol.

**References**      Miller WL, Portale AA. Vitamin D 1 alpha-hydroxylase. *Trends Endocrinol Metab* 2000;11:315–9 (Review).

# Hypophosphatasia

(also known as: phosphoethanolaminuria)

| | |
|---|---|
| **MIM** | 241500 (infantile), 241510 (childhood), 146300 (adult) |
| **Clinical features** | Clinical expression of hypophosphatasia is extremely variable, ranging from fetal death to mild osteopenia, and, in some cases, an absence of clinical symptoms. Dental abnormalities may be the only manifestation in some patients (odontohypophosphatasia). Other patients may show typical clinical and radiological signs of hypophosphatasia, but have normal serum levels of alkaline phosphatase (pseudohypophosphatasia). The infantile form is characterized by hypercalcemia, hypercalciuria, enlarged cranial sutures, craniosynostosis, delayed dentition, enlarged epiphyses, and prominent costochondral junctions. Rickets may be the only manifestation of hypophosphatasia in childhood. Adults show few symptoms despite osteopenia; metatarsal fractures may heal slowly and accelerated loss of teeth may occur. |
| **Primary tissue or gland affected** | Bones and teeth. |
| **Age of onset** | Perinatal, infancy, childhood, or adulthood. |
| **Epidemiology** | Sixty-five distinct mutations have been reported in approximately 70 unrelated patients from Europe (mostly France and Germany), North America, and Japan. Most patients are compound heterozygotes. |
| **Inheritance** | Autosomal recessive. |
| **Chromosomal location** | 1p36.1-p34 |
| **Genes** | *ALPL* (alkaline phosphatase liver-type; also known as *TNSALP* [tissue-nonspecific alkaline phosphatase]). |

| | |
|---|---|
| **Mutational spectrum** | Most are missense mutations. Splicing mutations, small deletions, frame-shift mutations, and a mutation affecting the transcription initiation site have all been described. |
| **Effect of mutation** | Mutations result in variable decreases in TNSALP activity. Pyrophosphate, an inhibitor of bone mineralization, is normally hydrolyzed by alkaline phosphatase. When alkaline phosphatase is deficient or absent, increased pyrophosphate levels interfere with normal bone mineralization. In pseudohypophosphatasia, substrate specificity or cellular localization of the enzyme is abnormal. |
| **Diagnosis** | Deficiency of TNSALP and increased urinary excretion of phosphoethanolamine in the presence of osteomalacia or rickets indicate a diagnosis of hypophosphatasia. When atypical radiologic or borderline biochemical findings are present, screening for *TNSALP* mutations may be helpful to confirm the diagnosis, and to exclude other skeletal disorders, such as osteogenesis imperfecta. |
| **Counseling issues** | Prenatal diagnosis should be offered to families affected with severe forms of hypophosphatasia in whom specific *TNSALP* mutations have been identified. |
| **Notes** | Medical treatment has not been effective. Surgical management of craniosynostosis in children and treatment of fractures in adults may be necessary. Intensive dental care is required to prevent tooth loss. |
| **References** | Mornet E. Hypophosphatasia: The mutations in the tissue-nonspecific alkaline phosphatase gene. *Hum Mutat* 2000;15:309–15 (Review). |
| | Zurutuza L, Muller F, Gibrat JF et al. Correlations of genotype and phenotype in hypophosphatasia. *Hum Mol Genet* 1999;8:1039–46. |

# Osteoporosis

(also known as: senile osteoporosis; postmenopausal osteoporosis)

| | |
|---|---|
| MIM | 166710 |

| | |
|---|---|
| Clinical features | Osteoporosis is a common medical problem, characterized by a reduction in bone mass and microarchitectural changes that increase the risk of fracture. Although particularly frequent in postmenopausal women, osteoporosis is also common in elderly men. Bone mass increases progressively during childhood and adolescence, reaching a peak between the ages of 25 and 35 years. After this age, bone mass decreases progressively. Estrogen deficiency, after natural or surgical menopause, causes a marked increase in the rate of bone loss. Hypogonadism in men causes a similar acceleration in bone loss. Twin and family studies have shown that up to 85% of the variance in bone density between individuals is genetically determined. Furthermore, other determinants of bone strength are genetically determined. These may include genetic effects on hip geometry and bone microarchitecture that can have a major impact on fracture risk, independent of bone mineral density. |
| Primary tissue or gland affected | Bone. |
| Age of onset | Adulthood. |
| Epidemiology | About 30% of postmenopausal women have osteoporosis. At age 50 years, the estimated life-time risk of hip fractures is 16% for women and 5% for men. The true prevalence of vertebral fractures is unknown, but may reach 16% in women and 6% in men after the age of 50 years. Many of these are asymptomatic. |
| Inheritance | Polygenic. |

| Chromosomal location | 17q21.31-q22, 7q22.1, 7q21.3, 7p21. |
|---|---|
| Genes | Genes that have been implicated by association and linkage studies include: *VDR* (vitamin D receptor), *COLIA1* (type 1 collagen), *TGFB* (transforming growth factor-β), *ER* (estrogen receptor), *IL-6* (interleukin-6), *APOE* (apolipoprotein E), and *BGP* (bone Gla protein; also known as osteocalcin). The relative importance of each is still unknown, and linkage studies suggest that other, as yet unidentified, genes may play critical roles. |
| Mutational spectrum | Osteoporosis is a polygenic disease. Common polymorphisms in a large number of genes may alter the disease risk in a given individual. |
| Effect of mutation | As a complex, polygenic disease, defects in any of many aspects of bone mineralization and strength can increase the risk of osteoporosis and osteoporotic fractures. These include defects in calcium absorption, mineralization, defective bone matrix formation, and structural changes that change stress levels at critical regions of the bone. Even seemingly unrelated factors such as balance and co-ordination may have important effects on the prevalence of fractures by increasing the risk of falling. |
| Diagnosis | The diagnosis of osteoporosis depends on the demonstration of bone density in a range that significantly increases fracture risk. Because the absolute value of bone density varies depending upon the technology used to measure it, the extent of bone loss is typically reported as the number of standard deviation (SD) units from the mean bone density of a young, healthy, adult control group studied using the same technology (T score). Individuals with bone mineral density values >2.5 SD below the mean for young healthy adults are considered to have osteoporosis. Bone density levels 1.0–2.5 SD below the mean indicate a diagnosis of osteopenia, an important risk factor for frank osteoporosis. |

**Counseling issues**  A major risk factor for osteoporosis is peak bone density during the third and fourth decades of life. Therefore, prevention of osteoporosis begins during adolescence and young adulthood. Untreated hypogonadism, vitamin D deficiency, inadequate calcium intake, and decreased physical activity are all associated with low peak bone density. Women with a family history of osteoporosis should have their bone density measured at the time of menopause, and changes should be monitored from then on. Because prevention is much more effective than treatment of established osteoporosis, efforts should be made to prevent accelerated bone loss by guaranteeing adequate calcium and vitamin D intake, and by hormone replacement therapy.

**Notes**  As with other polygenic diseases, such as type 2 diabetes, great efforts are being made to define the specific genetic risk factors. Although the genetics of polygenic diseases are of primarily academic interest at the present time, it is likely that the knowledge obtained from these studies will soon permit the tailoring of treatments based on the specific genetic risk factors present in a given individual. Furthermore, a better understanding of the pathophysiology of disease is likely to identify novel drug targets for the prevention and treatment of osteoporosis.

**References**  Ralston SH. Genetics of osteoporosis. *Rev Endocr Metab Disord* 2001;2:13–21 (Review).

# Osteogenesis Imperfecta

(includes: OI; brittle bone disease; OI tarda; OI with blue sclerae [type 1];
OI congenita, neonatal lethal form [type 2]; OI, progressively deforming, with normal
sclerae [type 3]; OI with normal sclerae [type 4])

| | |
|---|---|
| MIM | 166200 (type 1), 166210 (type 2), 259420 (type 3), 166220 (type 4) |
| Clinical features | OI is a heterogeneous syndrome in which the osteoblast produces an abnormal bone matrix that is unable to withstand normal mechanical loads. A high rate of bone turnover results from a compensatory increase in the number of osteoblasts and in osteoclast activity. Type 1 OI patients have a typical triangular-shaped face, blue sclerae, and sensorineural hearing loss beginning in adolescence to early adulthood. They may also develop neurologic symptoms, including hydrocephalus, paresthesias, and weakness. Patients may have had 5–15 bone fractures by puberty. The risk of fractures drops during puberty, but increases after the age of 50 years. Infants with the severe, lethal form (type 2) rarely survive beyond the first day of life due to multiple *in utero* fractures, a small thorax, and poorly developed thoracic muscles. Features of type 3 or 'severely deforming' OI include limb deformities, kyphoscoliosis, and significant growth retardation (final height may reach only 90–100 cm). Blue sclerae are noted only occasionally in type 3 patients. Type 4 patients have skeletal deformities that are intermediate in severity between those seen in types 1 and 3. Blue sclerae are usually not observed in type 4 patients. |
| Primary tissue or gland affected | Bone. |
| Other organs, tissues, or glands affected | Other connective tissue. |

| | |
|---|---|
| **Age of onset** | The lethal form of OI (type 2) may present during pregnancy when an ultrasound shows multiple intrauterine fractures in the fetus. At delivery, dismemberment of the neonate's head or an extremity may occur due to the defect in collagen synthesis. OI type 3 may present with fractures at birth and throughout infancy. In types 1 and 4, fractures are usually not present at birth, but typically occur in childhood. Type 4 patients may experience recurrent fractures after the age of 50 years in men, and after the menopause in women. |
| **Epidemiology** | OI is estimated to affect approximately 15 000 people in the United States. The true prevalence may be even greater, perhaps as high as 1:5000–10 000, since mild cases may not be recognized. Type 1 OI accounts for about 50% of all diagnosed cases; 5% have type 2; 20% have type 3; and 25% have type 4. The incidence of type 2 (severe) disease varies from 1:20 000–60 000. |
| **Inheritance** | Autosomal dominant, autosomal recessive. |
| **Chromosomal location** | 17q21.31-q22 (types 1, 2, 3, 4), 7q22.1 (types 1, 2, 3). |
| **Genes** | *COL1A1* ($\alpha$1 chain of type 1 collagen), *COL1A2* ($\alpha$2 chain of type 1 collagen). |
| **Mutational spectrum** | Over 250 mutations of the *COL1A1* and *COL1A2* genes have been described. Null alleles of *COL1A1* and *COL1A2* resulting from amino acid substitution, splicing defect, nonsense, and frame-shift mutations have been found in patients with type 1 disease. Multiexon deletions of the *COL1A2* gene have been reported in two patients with type 1B OI. Partial gene deletions encoding three exons, small helical deletions, and a single glycine substitution are associated with type 2 disease. Some cases of type 3 OI are caused by a 4-base pair deletion of *COL1A2*, resulting in a frame shift, while others result from the same mutations seen in severe OI. Mutations described in type 4 patients include a splicing abnormality and 15-amino acid deletion of COL1A2, a glycine point mutation, and a 9-base pair deletion. |

**Effect of mutation**     Null mutations in type 1 OI decrease the synthesis of type 1 collagen to about half. Mutations causing types 2 and 3 OI disrupt the integrity of the helical domain of collagen. In type 4 OI, the helical domain is disrupted or destabilized. In many cases, no correlation between disease severity and a specific mutation can be demonstrated. Patients with identical mutations sometimes manifest variable phenotypic expression of the disease. Approximately 25% of severe cases show no detectable collagen mutation.

**Diagnosis**     Decreased bone mineral density without major skeletal deformities in a child who has repeated fractures suggests a diagnosis of type 1 OI. A bone biopsy may be helpful to exclude other diagnoses and to define the severity of the disease. However, histology varies with the type of OI and the patient's age. Bone crystals (apatite) are small in cortical bone specimens from children and adolescents with type 2 OI, but crystal size increases in older patients. Tetracycline and demethylchlortetracycline may be given prior to bone biopsy in order to assist in analyzing rates of bone formation. Serum levels of alkaline phosphatase, osteocalcin, C-terminal propeptide of procollagen 1, and serum procollagen N-terminal propeptide reflect osteoblastic bone formation. Collagen breakdown may be measured by urinary excretion of collagen cross-link products, the *N*-telopeptide, or pyridinoline and deoxypyridinoline cross-links. Results of serum and urinary indicators of bone turnover and collagen breakdown are variable depending on the type and severity of OI and the age of the patient.

**Counseling issues**     Ultrasound can be used for prenatal diagnosis of severe OI, beginning at week 16 of pregnancy. DNA sampling from a chorionic villus biopsy may be helpful in families with a previously affected child. For severe OI, the risk of sibling recurrence is said to vary from 1% to 7%.

**Notes**     Child abuse is often mistakenly suspected in infants and young children in whom OI presents as multiple fractures. Attempts to improve bone mass and bone structure with calcitonin, calcium, fluoride, androgenic steroids, vitamin D, and growth hormone have been disappointing. Bone marrow transplantation has been performed in a small number of patients, but has not met with wide acceptance. Currently, administration of bisphosphonates, together with calcium and vitamin D supplementation, appears to be the treatment of choice, along with physical therapy and orthopedic surgery to correct skeletal deformities and facilitate weight bearing.

**References**

Glorieux FH. Osteogenesis imperfecta. A disease of the osteoblast. *Lancet* 2001;358:S45.

# Achondroplasia and Related Skeletal Dysplasias

(includes: hypochondroplasia; thanatophoric dysplasia (TD); severe achondroplasia with developmental delay and acanthosis nigricans [SADDAN])

| | |
|---|---|
| **MIM** | 100800 (achondroplasia), 146000 (hypochondroplasia), 187600 (TD) |
| **Clinical features** | Disproportionate short stature is a common feature of skeletal dysplasias. Characteristic features of achondroplasia include rhizomelic (proximal) short limbs, relative macrocephaly, midface hypoplasia, lumbar lordosis, trident hands, and incomplete extension of the elbow. Radiologic features include short tubular bones, 'squared-off' iliac wings, a narrow sacrosciatic notch, and distal reduction of the vertebral interpediculate distance. Skeletal changes and short stature in hypochondroplasia are milder. Limbs are short, with mild metaphyseal flaring, and lumbar lordosis may be present. TD is usually lethal in the neonatal period. Affected infants show micromelic shortening of the extremities, macrocephaly, platyspondyly, and a small thoracic cavity with short ribs. There are two phenotypic subtypes: in TD1, the femurs are curved; in TD2, straight femurs and a cloverleaf skull are seen. Patients with SADDAN dysplasia have extremely short stature, severe tibial bowing, marked developmental delay, hydrocephalus, seizures in infancy, and acanthosis nigricans. |
| **Primary tissue or gland affected** | Bone. |
| **Other organs, tissues, or glands affected** | Central nervous system and skin (SADDAN dysplasia). |
| **Age of onset** | Ultrasound examination in the second and third trimesters may detect skeletal abnormalities in the fetus. TD and homozygous achondroplasia are apparent immediately at birth. Patients with |

mild forms of hypochondroplasia may present during childhood for evaluation of short stature.

**Epidemiology**

Achondroplasia is the most common form of disproportionate short stature, with an incidence of 1:15 000–40 000 live births. Achondroplasia is sporadic in >90% of cases, and is associated with advanced paternal age. TD affects 1:60 000 births.

**Inheritance**

Autosomal dominant.

**Chromosomal location**

4p16.3

**Genes**

*FGFR3* (fibroblast growth factor receptor 3).

**Mutational spectrum**

Achondroplasia shows remarkable genetic homogeneity: >97% of patients have an arginine for glycine substitution (Gly380Arg) in the transmembrane domain of the FGFR3 protein. In sporadic cases, *FGFR3* mutations occur exclusively on the paternally derived chromosome. Hypochondroplasia shows a greater degree of clinical and genetic heterogeneity, including an Asn540Lys substitution in the proximal tyrosine kinase domain, and an Ile538Val substitution, also in the tyrosine kinase domain. In some families with a mild form of hypochondroplasia, no linkage to chromosome 4 has been found. TD resembles achondroplasia in its genetic homogeneity. TD1 results from a stop-codon mutation or missense mutation in the extracellular domain of the gene. TD2 is due to a Lys650Glu mutation in the tyrosine kinase 2 domain. A Lys650Met mutation was reported in three unrelated patients with SADDAN dysplasia, and in two patients with TD1.

**Effect of mutation**

The achondroplasia family of disorders results from *FGFR3* gain of function mutations that cause constitutive activation of a normally negative growth-control pathway. Activation of the transcription factor STAT1 (signal transducer and activator of transcription) and expression of p21(WAF/CIP1), a cell-cycle inhibitor, may explain the inhibition of chondrocyte proliferation observed in this group of skeletal dysplasias. There is a strong correlation between phenotype and genotype for achondroplasia and TD. Mutations causing TD activate FGFR3 more strongly than the mutations that result in achondroplasia.

**Diagnosis**

Accurate diagnosis of skeletal dysplasia cannot usually be made on the basis of prenatal ultrasound prior to the third trimester. Because of the strong phenotype–genotype correlation in achondroplasia and the high frequency of a specific *FGFR3* mutation, DNA analysis can provide the basis for prenatal diagnosis. Mutational analysis is particularly important for prenatal diagnosis of the lethal form of homozygous achondroplasia. Because there are many other lethal types of skeletal dysplasia that are not accounted for by *FGFR3* mutations, the diagnosis of infants with short limbs and a small thorax detected on prenatal ultrasound, in the absence of a family history, will need to wait until after birth. At this point, appropriate radiologic and, if necessary, histologic studies can be performed. Disproportionately short limbs should be suspected if the arms do not reach the midpelvis in infancy or the upper thigh after infancy. Measurements of arm span and sitting height (upper:lower body segment ratio) are needed to detect mild degrees of disproportionate short stature. Rhizomelic dysplasia involves shortening of the proximal segments (upper arms and legs). The middle segments (forearms and lower legs) are short in mesomelic dysplasia, while acromelic shortening affects the hands and feet. Campomelic dysplasia refers to bowing of the legs. A radiologic evaluation (skeletal survey) should be performed before puberty, since evaluation of the epiphyses and metaphyses is difficult or impossible after epiphyseal closure. Because most forms of skeletal dysplasia are symmetric, evaluation of the extremities on one side only is usually adequate. For a fetus or neonate, anteroposterior (AP) and lateral X-rays of the whole body should be performed. For older patients, recommended X-rays include AP and lateral views of the skull and spine; AP views of the thorax (ribs), pelvis, left hand, and the long bones on one side only; and a lateral view of the knee (patella). Histologic and electron microscopic studies of bone and cartilage are usually performed only at the time of autopsy or surgical intervention because of an orthopedic complication.

**Counseling issues**

Patients with achondroplasia are fertile. When both parents are heterozygous for a mutation, the risk to their offspring of lethal, homozygous achondroplasia is 25%. In contrast to achondroplasia, the genotype and phenotype are not strongly correlated in hypochondroplasia.

**Notes**

Attempts to treat achondroplasia patients with growth hormone have resulted in some increase in growth rate during the first year, with a lesser response in the second year of treatment. No information regarding final height is available at this time. The overall prevalence of skeletal dysplasia is approximately 1:4000 births. The nearly 200 distinct clinical entities can be divided into the osteodysplasias (e.g. osteogenesis imperfecta) and chondrodysplasias (e.g. achondroplasia).

**References**

Mortier GR, Weis M, Nuytinck L et al. Report of five novel and one recurrent COL2A1 mutations with analysis of genotype–phenotype correlation in patients with a lethal type II collagen disorder. *J Med Genet* 2000;37:263–71.

Vajo Z, Francomano CA, Wilkin DJ. The molecular and genetic basis of fibroblast growth factor receptor 3 disorders: The achondroplasia family of skeletal dysplasias, Muenke craniosynostosis, and Crouzon syndrome with acanthosis nigricans. *Endocr Rev* 2000;21:23–39 (Review).

http://www.LPAonline.org/resources_library.html

# 5. Disorders of Glucose Metabolism

# Syndromes and diseases associated with hyperglycemia

This chapter opens with a discussion of two common polygenic forms of diabetes, types 1 and 2 diabetes mellitus. The precise genetic etiology of these is unknown, but because of their very high and increasing prevalence, and because both have very strong genetic components, practicing internists, pediatricians, and endocrinologists are often asked to explain the clinical relevance of genetic predisposition to these common and heterogeneous syndromes. These two entries are followed by various rare monogenic forms of hyperglycemia.

# Syndromes and diseases associated with hypoglycemia

Primary hypoglycemia can be the presenting symptom of a variety of genetic abnormalities. We first discuss familial hyperinsulinism of infancy, a heterogeneous syndrome that can be caused by mutations in any of at least four β-cell-specific genes (see **Figure 1**). This is followed by a discussion of other defects in carbohydrate or fat metabolism. Although these are not hormonal in pathophysiology, they frequently present with hypoglycemia and may therefore be referred to an endocrinologist for initial diagnosis.

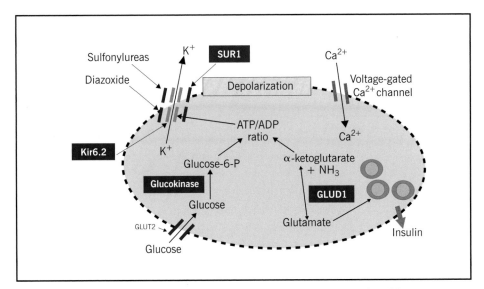

**Figure 1.** Hyperinsulinemic hypoglycemia can be caused by mutations in at least four different genes, highlighted in purple. SUR1 and Kir6.2 form the KATP channel, a hetero-octamer. Under normal resting conditions the channel is open, thus maintaining the membrane in a hyperpolarized condition. Metabolism of glucose causes an increased intracellular ATP/ADP ratio, resulting in closure of the channel, depolarization of the membrane, opening of the voltage-gated $Ca^{2+}$ channels, and, finally, insulin secretion. The mechanism by which glutamate dehydrogenase (GLUD1) causes hyperinsulinism is still controversial. GLUT2: glucose transporter 2; Kir6.2: inward rectifying potassium channel; SUR1: sulfonylurea receptor 1.

# Type 1 Diabetes Mellitus

**(also known as: T1DM; insulin-dependent diabetes mellitus)**

| | |
|---|---|
| **MIM** | 222100 |

**Clinical features**  Symptoms usually appear over the course of days or weeks in nonobese children and adolescents. The onset of T1DM appears to be more gradual in adults. Typical symptoms include polyuria, polydipsia, polyphagia, pruritus, and weight loss. Ketoacidosis develops when the diagnosis and initiation of insulin treatment are delayed. Some patients experience a 'honeymoon' period, lasting several weeks or months, during which insulin requirements fall dramatically. Late microvascular (retinopathy and nephropathy), macrovascular, and neuropathic complications develop with increased frequency and severity in patients with poor glycemic control.

**Primary tissue or gland affected**  Pancreatic islet β cells.

**Age of onset**  The incidence of T1DM increases progressively from infancy, reaches a peak at puberty, and declines through early adulthood. With the increasing use of laboratory tests to measure autoimmunity, more cases of T1DM are being found in adult patients who would previously have been diagnosed with T2DM.

**Epidemiology**  The risk of developing T1DM is approximately 0.3%–0.4% in a general population with no family history of diabetes. T1DM is more prevalent in Finland, Scandinavia, Scotland, and Sardinia, less prevalent in southern Europe and the Middle East, and uncommon in Asia. The prevalence of genes that confer susceptibility to, or protect against, the development of T1DM varies among different ethnic groups, and accounts for some of the variability seen in regional frequency data. Seasonal variation in the onset of diabetes has been noted in a wide range of geographic locations and ethnic groups; the incidence is highest in late fall and early winter.

| **Inheritance** | The precise mode of inheritance for T1DM is unknown. Some findings are consistent with dominant inheritance, but the rare frequency of occurrence in first-degree relatives does not support classical dominant heredity. The high frequency of two shared haplotypes in concordant siblings suggests recessive inheritance. Other findings cast doubt on the concept of recessive inheritance, namely that risk for diabetes is not increased in subjects who are homozygotic for the human leukocyte antigen (HLA)-DR3 or HLA-DR4 allele, while heterozygosity for specific DR3/DR4 alleles does confer increased risk. T1DM may best be described as a complex genetic disorder in which multiple susceptibility genes interact with environmental factors to initiate or modify autoimmune β-cell destruction. |
|---|---|
| **Chromosomal location** | 6p21.3, 11p15.5 |
| **Genes** | The IDDM1 locus on chromosome 6 encompasses the HLA region within the major histocompatibility complex (MHC). Approximately 40% of the genetic susceptibility to T1DM can be explained by the IDDM1 locus. At least three genes (HLA-DRB1, HLA-DQA1, and HLA-DQB1), located in the class II region, play a major role in determining susceptibility to T1DM. The HLA-DQ alleles show the strongest association with T1DM. The IDDM2 locus maps to a variable number of tandem repeats (VNTR) region located 596-bp upstream of the insulin gene translational start site on chromosome 11. At least 18 other chromosomal regions have been identified as possible susceptibility genes for T1DM in various populations and ethnic groups. The understanding of the genetics of T1DM is expected to expand exponentially in the coming years. |
| **Mutational spectrum** | IDDM1: DQA1*0501-DQB1*0201 and DQA1*0301-DQB1*0302 encode the DQ2 and DQ8 molecules, respectively; heterozygosity for DQ2 and DQ8 carries the highest risk for T1DM. The HLA-DR alleles, DRB1*03 and DRB1*04, are also associated with increased risk. By contrast, DRB1*0403 and DRB1*0406 confer protection from diabetes. IDDM2: Alleles of VNTR are classified according to |

the number of repeats of a 14- to 15-bp sequence: class 1 (26–63 repeats), class 2 (64–139 repeats), and class 3 (140–210 repeats). Class 1 homozygosity is associated with the highest risk of T1DM; class 3 alleles confer dominant protection.

**Effect of mutation**

MHC class II molecules are expressed on the cell surface of macrophages and other antigen-presenting cells. A peptide groove, formed by the α and β chains of these MHC molecules, binds antigens involved in the pathogenesis of T1DM and allows these antigens to activate receptors on T-cells. These receptors, in turn, carry out the autoimmune destruction of islet β cells. Substitutions in the amino acid composition of the MHC α and β chains affect their ability to bind specific antigens, and may thereby increase or decrease susceptibility to developing T1DM. Mutations leading to the substitution of any amino acid other than aspartate at position 57 on the DQ β chain result in increased risk for T1DM. The exact functional significance of the insulin gene VNTR in the pathogenesis of T1DM is unknown. VNTR may regulate the expression of the insulin and *IGF2* genes. The degree of insulin mRNA expression in the fetal thymus could modulate autoimmunity or tolerance. Placental *IGF2* expression could affect intrauterine growth and birth weight, which may affect risk for T1DM.

**Diagnosis**

In the presence of typical clinical symptoms, a random blood glucose >200 mg/dL is sufficient to diagnose diabetes mellitus. When classical symptoms are absent, a fasting blood glucose >126 mg/dL on at least two separate occasions, or blood glucose >200 mg/dL on oral glucose tolerance testing, will identify the patient with diabetes mellitus. Assignment of a specific diagnosis of T1DM may require further observation of the clinical course, including insulin requirements, tendency to develop ketosis, and the presence of immune markers associated with T1DM. Differential diagnoses include T2DM, which is recognized with increasing frequency in children and adolescents, maturity-onset diabetes of the young, and secondary forms of diabetes. Islet cell antibodies can be detected in approximately 80% of patients with new-onset T1DM. Antibodies have been characterized that are directed against specific islet cell antigens, including insulin, glutamic acid decarboxylase (GAD65 and GAD67), and ICA512 – a secretory granule protein with a tyrosine phosphatase-like domain.

**Counseling issues**

In a North American population, the risk of T1DM in the offspring of a T1DM father is estimated to be about 5%, compared with a 2% risk for the offspring of a T1DM mother. Although the overall prevalence of T1DM varies among different populations, the risk of TIDM for the offspring of a diabetic father is 2–3 times greater than the risk for maternal offspring. The risk to siblings is estimated to be 5%, and for identical twins 30%–50%. The risk for siblings sharing no, one, or two HLA haplotypes is approximately 2%, 5%, and 13%, respectively. The sharing of two haplotypes with a DR3/DR4 proband confers the highest sibling risk of 19%. The risk of developing T1DM is increased among relatives with positive measures of autoimmunity, and is highest among relatives who test positive for multiple antibodies. Interventions aimed at blocking the autoimmune destruction of β cells in high-risk individuals have included the use of azathioprine, cyclosporine, nicotinamide, low-dose parenteral and oral insulin, and diet modifications such as avoidance of cow's milk. While the initial results from high-risk patients treated with low-dose insulin appeared to be encouraging, a large multicenter study (the Diabetes Prevention Trial [DPT]-1) showed no difference in the rate of developing diabetes in the low-dose insulin-treated group compared with untreated controls. The rate of developing diabetes was 60% in each group of these high-risk subjects. Although not successful in preventing the disease, this study confirms that it is possible to identify subjects at high risk for the development of diabetes using a combination of genetic, immunologic, and metabolic (insulin responses to intravenous glucose) markers. A trial of oral insulin in patients at moderate risk for developing diabetes is currently in progress. To date, none of the intervention trials have produced and maintained long-term, insulin-free remissions. Current studies include the

DPT-1 oral insulin trial, the European Nicotinamide Diabetes Intervention Trial, and immunotherapy using interleukin-2. Other recent studies have involved DiaPep277, an immunomodulatory peptide derived from heat-shock protein.

**Notes**

The terms 'juvenile diabetes' or 'juvenile-onset diabetes mellitus' should no longer be used, since T1DM may be diagnosed in adults as well as children.

**References**

Bennett ST, Todd JA. Human type 1 diabetes and the insulin gene: Principles of mapping polygenes. *Annu Rev Genet* 1996;30:343–70 (Review).

Kelly MA, Mijovic CH, Barnett AH. Genetics of type 1 diabetes. *Best Pract Res Clin Endocrinol Metab* 2001;15:279–91 (Review).

# Type 2 Diabetes Mellitus

(also known as: T2DM; maturity-onset diabetes [MOD]; noninsulin-dependent diabetes mellitus [NIDDM])

| | |
|---|---|
| **MIM** | 125853 |

**Clinical features**    T2DM is typically progressive, beginning with mild postprandial and/or fasting hyperglycemia that can be controlled by lifestyle intervention alone (diet and exercise). Over time, this treatment becomes insufficient and oral medication must be added. This includes classes of drugs that increase the sensitivity of the body to endogenous insulin (such as metformin and the glitazones), and drugs that stimulate endogenous insulin secretion (such as the sulfonylureas and meglitinides). With progression of the disease, multidrug treatment is typically required. Frequently, oral medication becomes inadequate to maintain euglycemia, and exogenous insulin is required. T2DM can be differentiated clinically from T1DM since the latter is associated with complete β-cell failure and the presence of antibodies to insulin and β-cell components, indicating an autoimmune process leading to β-cell destruction. However, in some cases, the clinical distinction can be difficult because adults with slowly progressing T1DM, so-called latent autoimmune diabetes in the adult (LADA), may have antibody titers that are below the level of detection in common clinical assays. The primary pathology of T2DM is hyperglycemia, which in most cases can be relatively easily managed. However, long-term complications, including microvascular, macrovascular, and neuropathic complications, result in most of the morbidity and mortality from the disease. Recent studies have convincingly demonstrated that optimal glycemic control is associated with a marked reduction in the incidence of these complications. A major effort on the part of the patient and the treatment team is required to achieve the level of glycemic control required.

| | |
|---|---|
| **Primary tissue or gland affected** | Pancreatic β cells, muscle, fat, liver. |
| **Age of onset** | The onset of clinical disease is typically in patients >40 years old. However, over the last 20 years a marked increase in the incidence of adolescent-onset disease has been noted. This is thought to be caused by changes in lifestyle, including increased caloric intake with decreasing levels of physical activity. |
| **Epidemiology** | The incidence of T2DM is dependent on the combined level of genetic and environmental risk factors. Both appear to be age related. In some populations, such as the Pima Indians, a very high level of both results in an extremely high overall prevalence (about 50% in adults). In the USA and western Europe, the prevalence is >10% after the age of 60 years. Because of changes in lifestyle and life expectancy throughout the world, it is estimated that the prevalence of disease will increase by almost 50% between 2000 and 2010. |
| **Inheritance** | T2DM is a polygenic disease. The currently accepted genetic model of T2DM states that variations in any of many different genes may increase the risk of developing T2DM. The net genetic risk of any particular individual is thought to be defined by the number and severity of inherited genetic risk factors. |
| **Chromosomal location** | 2q37.3 (NIDDM1). Other loci have been identified by linkage analysis and association studies; however, the clinical significance of these is still unknown. |
| **Genes** | Calpain-10 has been called NIDDM1 to signify that it is the first gene identified by positional cloning to be associated with the common, polygenic form of T2DM. Genetic variants in a large number of other candidate genes, including SUR1, Kir6.2, PPAR-γ (peroxisome proliferator-activated receptor-γ), IRS1 (insulin receptor substrate 1), and others have been associated with T2DM in some populations. The significance of all of these findings is still controversial. |

| | |
|---|---|
| **Mutational spectrum** | An association between an intronic point polymorphism (UCSNP-43) in calpain-10 and diabetes-related phenotypes has been reported, though its significance is very controversial. This polymorphism may affect gene expression in skeletal muscle. |
| **Effect of mutation** | T2DM is characterized by variable degrees of resistance to insulin action in association with defective β-cell function. Although the primary defect responsible for disease is still controversial, it is now clear that some form of β-cell dysfunction must be present for frank hyperglycemia to appear. |
| **Diagnosis** | The utility of the oral glucose tolerance test has been questioned because of poor reproducibility, and this test is now rarely used for clinical diagnosis. New diagnostic criteria are primarily based on fasting blood glucose levels. A fasting blood glucose level <110 mg/dL (6.1 mmol/L) is considered normal. Fasting levels >126 mg/dL (7 mmol/L) are considered diagnostic of diabetes. Intermediate levels are termed 'impaired fasting glucose'. About 40% of patients with impaired fasting glucose will go on to develop diabetes within 5 years. A random blood glucose level >200 mg/dL (11.1 mmol/L) is also considered diagnostic of diabetes. Because of the day-to-day variability of glucose levels, tests should generally be repeated for confirmation. |
| **Counseling issues** | The genetic risk of T2DM cannot be accurately defined at present, and must be based on family history. The overall life-time risk varies from <5% in patients with no family history at all, to nearly 100% in identical twins of affected probands. Individuals with one first-degree relative with T2DM have a 10%–30% chance of developing the disease. Careful avoidance of all environmental risk factors, particularly obesity and lack of physical exercise, will greatly reduce the risk of developing clinical disease. Furthermore, periodic monitoring of glucose levels and early, intensive management of even mild hypoglycemia are likely to improve the overall prognosis. |

**Notes**

The definition and clinical significance of 'impaired fasting glucose' are still controversial. Recent data have suggested that the older criterion of 'impaired glucose tolerance', defined as blood glucose 140–200 mg/dL (7.8–11.1 mmol/L) 2 hours after a 75 g glucose load, may be a better predictor of long-term cardiovascular risk than the new criteria based only on fasting glucose levels.

**References**

Froguel P, Velho G. Genetic determinants of type 2 diabetes. *Recent Prog Horm Res* 2001;56:91–105 (Review).

# Maturity-onset Diabetes of the Young

(also known as: MODY)

| | |
|---|---|
| **MIM** | 125850 (MODY1), 125851 (MODY2), 600496 (MODY3), 606392 (MODY4), 604284 (MODY5), 606394 (MODY6) |
| **Clinical features** | Mutations in any of the genes listed below will produce a clinical picture similar to that of type 2 diabetes; however, the degree of severity varies between the subgroups of this syndrome. MODY2, which is caused by a mutation in the glucokinase gene, is typically mild and rarely has long-term complications, with the primary defect being a change in the glucose set point. A patient with severe neonatal diabetes due to homozygosity for a glucokinase mutation has been reported. MODY1 and MODY3 are phenotypically similar, since they both cause defects in the same pathway of transcriptional regulation. Patients with mutations in these genes typically have severe disease with clear insulin deficiency and frequent long-term complications. MODY4, MODY5, and MODY6 appear to be associated with mild disease and rarely have complications; however, only a few individuals with these mutations have been described. Despite remarkable advances in the understanding of the genetics of this syndrome, the genetic etiology of 10%–20% of cases is still unknown. |
| **Primary tissue or gland affected** | Pancreatic β cells. |
| **Age of onset** | Variable. MODY2 typically presents in childhood, whereas MODY1, 3, and 5 present in the postpubertal period, and MODY4 in early adulthood. |
| **Epidemiology** | Recent studies suggest that 2%–5% of type 2 diabetic patients may in fact have MODY. In a recent study of a large UK cohort of MODY patients, 63% had hepatic nuclear factor 1α mutations and 20% had glucokinase mutations. The relative importance of these two subtypes |

may vary in different populations. MODY1, 4, 5, and 6 are very rare in all populations studied. Approximately 10%–20% of patients appear to have mutations in other, as yet unidentified, genes.

**Inheritance** Autosomal dominant.

**Chromosomal location** 20q12-q13.1 (MODY1), 7p15-p13 (MODY2), 12q24.2 (MODY3), 13q12.1 (MODY4), 17cen-q21.3 (MODY5), 2q32 (MODY6).

**Genes** *HNF4A* (hepatic nuclear factor 4α [MODY1]), *GCK* (glucokinase [MODY2]), *HNF1A* (hepatic nuclear factor 1α [MODY3]), *IPF1* (insulin promoter factor 1 [MODY4]), *HNF1B* (hepatic nuclear factor 1β [MODY5]), *NEUROD1* (neurogenic differentiation factor 1 [MODY6]). See **Figure 2**.

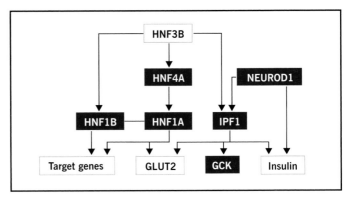

**Figure 2.** Transcription factors known to be important in β-cell development and insulin secretion. Genes associated with various forms of maturity-onset diabetes of the young (MODY) are highlighted in purple. GCK: glucokinase (MODY2); GLUT2: glucose transporter 2; HNF1A: hepatic nuclear factor 1α (MODY3); HNF1B: hepatic nuclear factor 1β (MODY5); HNF4A: hepatic nuclear factor 4α (MODY1); IPF1: insulin promoter factor 1 (MODY4); NEUROD1: neurogenic differentiation factor 1 (MODY6).

**Mutational spectrum** A large number of mutations have been found, with little correlation between the site of the mutation and the clinical presentation.

| **Effect of mutation** | Mutations can cause decreased protein function by at least two mechanisms. The first is a gene-dosage effect, meaning that two functioning alleles are required to produce the necessary amount of gene product; if one is defective, insufficient protein is produced. In the second mechanism, the abnormal product exerts a dominant-negative effect, meaning that it binds to the normal protein produced by the wild-type allele and interferes with its function. At least some HNF1$\alpha$ mutations appear to exert their effects by this mechanism. In the case of GCK, mutations decrease the affinity of the enzyme for glucose, thus increasing the glucose level at which insulin secretion is stimulated. The $\beta$ cells in heterozygous individuals coexpress wild-type and mutant GCK, resulting in a glucose set-point that is intermediate between that expected from normal and homozygous mutant $\beta$ cells. |
|---|---|
| **Diagnosis** | Diagnosis is based on the identification of relatively mild diabetes in a young individual, typically <25 years old, with a family history suggestive of autosomal dominant disease. If insulin therapy is initiated when hyperglycemia is diagnosed in children, these cases may be missed. Pediatric endocrinologists must keep the possibility of MODY in mind when presented with a child with an apparently mild glucose abnormality. At the time of presentation, it may be difficult to differentiate between MODY and very early type 1 diabetes with partial autoimmune $\beta$-cell destruction. Measuring anti-islet cell antibodies, antiglutamic acid decarboxylase antibodies, and insulin autoantibodies may help to make the distinction. However, patients with slowly progressive type 1 diabetes may have antibody levels that are below the level of detection of commonly used assays. In some cases, a definitive clinical diagnosis may not be possible at the time of presentation. Persistence of stable $\beta$-cell function for more than 1–2 years after the initial diagnosis is strong evidence against type 1 diabetes. This can be determined either by serial measurements of stimulated C-peptide concentrations, or clinically by documenting adequate glycemic control without the need for exogenous insulin. |

| | |
|---|---|
| **Counseling issues** | This is a treatable autosomal dominant disease with high penetrance. Patients at risk should be tested regularly and treated at the first sign of glucose intolerance. |
| **Notes** | About 10%–20% of MODY patients do not appear to have mutations in any of the five genes described here. It is therefore likely that more MODY genes will be identified in the future. |
| **References** | Fajans SS, Bell GI, Polonsky KS. Molecular mechanisms and clinical pathophysiology of maturity-onset diabetes of the young. *N Engl J Med* 2001;345:971–80 (Review). |
| | Froguel P, Velho G. Molecular genetics of maturity-onset diabetes of the young. *Trends Endocrinol Metab* 1999;10:142–6. |

# Wolfram Syndrome

(also known as: diabetes insipidus and diabetes mellitus with optic atrophy and deafness [DIDMOAD])

| | |
|---|---|
| **MIM** | 222300, 606201, 604928, 598500 |
| **Clinical features** | Diabetes mellitus (insulin requiring) is usually the first abnormality detected. This is followed by diverse neurologic symptoms, including central diabetes insipidus, optic atrophy, sensorineural hearing loss, autonomic dysfunction, nystagmus, mental retardation, seizures, and brain atrophy on magnetic resonance imaging. Other abnormalities that may be present include hydronephrosis, neutropenia, thrombocytopenia, megaloblastic anemia, sideroblastic anemia, and psychiatric illness. |
| **Primary tissue or gland affected** | Pancreatic β cells. |
| **Other organs, tissues, or glands affected** | Posterior pituitary. |
| **Age of onset** | Childhood, adolescence, or early adulthood. |
| **Epidemiology** | Rare. |
| **Inheritance** | Autosomal recessive, maternal (mitochondrial). |
| **Chromosomal location** | 4p16.1, 4q22-q24, mitochondrial. |
| **Genes** | *WFS1* (wolframin 1), *WFS2* (wolframin 2), *MtDNA* (mitochondrial DNA). |

| | |
|---|---|
| **Mutational spectrum** | More than 20 missense, nonsense, deletion, and insertion *WFS1* gene mutations have been associated with Wolfram syndrome. The *WFS2* gene has not yet been identified, though the presence of a disease-associated gene in this location is suggested by linkage analysis. Several mitochondrial mutations have been associated with this syndrome, but the specific mechanism by which these cause disease is still unknown. |
| **Effect of mutation** | In patients with *WFS1* mutations, no functional gene product is produced. |
| **Diagnosis** | Clinical diagnosis is based on the identification of primary clinical symptoms: diabetes mellitus and bilateral progressive optic atrophy. No other clinical signs are required to make the clinical diagnosis. The diagnosis may be confirmed by identification of mutations in the *WFS1* gene. |
| **Counseling issues** | It is crucial to document the genetic cause of Wolfram syndrome in a specific family, since the disease can be maternally inherited or transmitted as an autosomal recessive trait. The genetic heterogeneity of this clinical syndrome has made genetic counseling even more complex if a causative mutation cannot be identified. |
| **Notes** | Heterozygous carriers of *WFS1* mutations may be predisposed to psychiatric illness. The clinical syndrome appears to be genetically heterogeneous: in some families, linkage to 4q22-24 has been found (*WFS2*), and in others mtDNA mutations have been found. The cellular function of WFS is not yet known. However, it is an endoplasmic reticulum (ER) resident protein, and sequence similarity to other ER proteins suggests that it may have a role in degrading malfolded ER proteins, thus minimizing ER stress. Lack of WFS may place cells at risk of apoptosis secondary to ER stress. |

**References**

Inoue H, Tanizawa Y, Wasson J et al. A gene encoding a transmembrane protein is mutated in patients with diabetes mellitus and optic atrophy (Wolfram syndrome). *Nat Genet* 1998;20:143–8.

Rotig A, Cormier V, Chatelain P et al. Deletion of mitochondrial DNA in a case of early-onset diabetes mellitus, optic atrophy, and deafness (Wolfram syndrome, MIM 222300). *J Clin Invest* 1993;91:1095–8.

# Diabetes–Deafness Syndrome

(also known as: maternally inherited diabetes and deafness [MIDD]; Ballinger–Wallace syndrome; noninsulin-dependent diabetes mellitus with deafness)

| | |
|---|---|
| MIM | 520000 |
| Clinical features | Most patients have a normal or low body-mass index. Approximately 75% of patients have a history of maternal diabetes. At the time of diagnosis, the diabetes is usually clinically similar to that of classical type 2 diabetes. In most patients, hyperglycemia can initially be controlled with oral agents, though the β-cell dysfunction will progress in many, and insulin therapy may be required after about 10 years of disease. Sensorineural hearing loss precedes the onset of diabetes and is present in approximately half of the patients. Vestibular function may be impaired. Most patients have a specific retinal lesion known as macular pattern dystrophy (see **Figure 3**), and about half of the patients have myopathy. Cardiomyopathy and neuropsychiatric symptoms are also common. Mitochondrial disease may predispose to nephropathy, which is particularly common in this syndrome. |
| Primary tissue or gland affected | Pancreatic β cells. |
| Age of onset | In a recent series, the mean age of diagnosis was approximately 40 years, with about half of the patients diagnosed before the age of 35 years. |
| Epidemiology | It is estimated that 0.5%–2.8% of patients with type 2 diabetes may have this syndrome. |
| Inheritance | Maternal (mitochondrial). |
| Chromosomal location | Mitochondrial (mt)DNA |

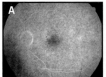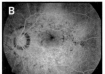

**Figure 3.** Early macular pattern dystrophy in a patient with maternally inherited diabetes–deafness syndrome. (**A**) Fluorescein angiography shows small pigmented deposits in the macula. (**B**) Fluorescein angiography shows mild atrophy of the retinal pigment epithelium at the posterior pole, combined with peripapillas and perimacular subretinal pigmented deposits. Reproduced with permission from the American College of Physicians (*Ann Intern Med* 2001;134:721–8).

| | |
|---|---|
| **Genes** | Mt-tRNA Leu-UUR (mitochondrial DNA mutation). |
| **Mutational spectrum** | The syndrome is caused by a point mutation at position 3243 of mtDNA. |
| **Effect of mutation** | Mitochondrial metabolism is impaired, resulting in decreased oxidative phosphorylation and cellular damage. In β cells, glucose sensing is impaired due to a decrease in ATP response to glucose metabolism. |
| **Diagnosis** | The clinical diagnosis is suspected whenever diabetes and sensorineural deafness coexist. Evidence of maternal inheritance, and the finding of associated neurologic symptoms and myopathy, make the diagnosis more certain. However, a definitive diagnosis can only be made by identifying the specific mutation in mtDNA. |
| **Counseling issues** | Inheritance is strictly maternal and severity is variable. Affected males can be assured that they will not transmit the disease to their offspring. |
| **Notes** | There is significant clinical similarity between several syndromes caused by mtDNA mutations, including the Wolfram (MIM 222300) and MELAS (mitochondrial myopathy, encephalopathy, lactic acidosis, and stroke-like syndrome; MIM 540000) syndromes. |

**References**

Guillausseau PJ, Massin P, Dubois-LaForgue D et al. Maternally inherited diabetes and deafness: a multicenter study. *Ann Intern Med* 2001;134:721–8.

Velho G, Byrne MM, Clement K et al. Clinical phenotypes, insulin secretion, and insulin sensitivity in kindreds with maternally inherited diabetes and deafness due to mitochondrial tRNALeu(UUR) gene mutation. *Diabetes* 1996;45:478–87.

# Alstrom's Syndrome

| | |
|---|---|
| **MIM** | 203800 |

| | |
|---|---|
| **Clinical features** | Alstrom's syndrome is characterized by blindness, deafness, obesity, and type 2 diabetes mellitus. Although similar to Bardet–Biedl syndrome (MIM 209900), Alstrom's syndrome does not typically include mental defects, polydactyly, or hypogonadism. The retinal lesion causes nystagmus, followed by progressive loss of central vision, differentiating this syndrome from most other pigmented retinopathies, in which the peripheral vision is lost first. Patients are typically tall during early childhood, with advanced bone age ultimately resulting in a decreased final height. Dilated cardiomyopathy is frequently seen during the first months of life. Diabetes is preceded by insulin resistance, hypertriglyceridemia, and low high-density lipoprotein cholesterol. Complications that are usually associated with diabetes, such as accelerated atherosclerosis, have been reported in older patients. Hyperuricemia and progressive renal failure have been reported in the Acadian kindred. Diabetes insipidus, hypothyroidism, male hypogonadism, asthma, liver disease, and cardiac fibrosis have been reported in cases of Alstrom's syndrome. However, the occurrence of these is variable, and it is not at all certain that they are truly part of the syndrome. |

| | |
|---|---|
| **Primary tissue or gland affected** | Pancreatic $\beta$ cells. |

| | |
|---|---|
| **Other organs, tissues, or glands affected** | Eyes, ears, heart. |

| | |
|---|---|
| **Age of onset** | Nystagmus, retinal dystrophy, hearing loss, obesity, and acanthosis nigricans appear during early childhood. Although obesity has been reported at as early as 6 months of age, infantile cardiomyopathy may be the earliest clinical sign. Type 2 diabetes develops after adolescence. |

| | |
|---|---|
| **Epidemiology** | This syndrome is rare, with only 90 families reported in the world literature between 1959 and 2002. Large multiplex kindreds have been reported in the Acadian population. |
| **Inheritance** | Autosomal recessive. |
| **Chromosomal location** | 2p13 |
| **Genes** | *ALMS1* (Alstrom syndrome gene). |
| **Mutational spectrum** | Frame-shift and nonsense mutations have been found in several families. One family was reported with a balanced, reciprocal translocation resulting in disruption of the *ALMS1* gene. Identification of the precise break site on chromosome 2p13 led to the discovery of the genetic defect of this syndrome. |
| **Effect of mutation** | These mutations clearly prevent the production of any functional protein. However, the function of the *ALMS1* gene product, which is ubiquitously expressed, is not known. |
| **Diagnosis** | Clinical diagnosis may be difficult during childhood, before the advent of overt diabetes. Differentiation between this and Bardet–Biedl syndrome may be difficult, and is based on a lack of polydactyly, and normal mental development. Infantile cardiomyopathy appears to be very common, and may aid in clinical diagnosis during early childhood. The gene responsible for Alstrom's syndrome was identified in 2002, and an early, definitive molecular diagnosis is therefore now possible. |
| **Counseling issues** | Since the genetic etiology of this syndrome is now known, prenatal diagnosis and genetic counseling are possible, once a specific mutation (or mutations) has been identified in a family. |

**References**

Collin GB, Marshall JD, Ikeda A et al. Mutations in ALMS1 cause obesity, type 2 diabetes and neurosensory degeneration in Alstrom syndrome. *Nat Genet* 2002;31:74–8.

Hearn T, Renforth GL, Spalluto C et al. Mutation of ALMS1, a large gene with a tandem repeat encoding 47 amino acids, causes Alstrom syndrome. *Nat Genet* 2002;31:79–83.

Marshall JD, Ludman MD, Shea SE et al. Genealogy, natural history, and phenotype of Alstrom syndrome in a large Acadian kindred and three additional families. *Am J Med Genet* 1997;73:150–61 (Review).

# Diabetes Mellitus, Insulin Resistant, with Acanthosis Nigricans and Hypertension

| | |
|---|---|
| **MIM** | 604367 |
| **Clinical features** | Patients present during the second or third decade of life with type 2-like diabetes, severe insulin resistance, hypertension that is difficult to control, and acanthosis nigricans. Women present with primary or secondary amenorrhea. Common polymorphisms in the gene responsible for this disease may be associated with severe obesity, and with the more prevalent phenotype of type 2 diabetes mellitus. |
| **Primary tissue or gland affected** | Pancreatic β cells. |
| **Age of onset** | The onset of diabetes is during the second or third decade of life. |
| **Epidemiology** | Although the clinical syndrome is not rare, mutations in the affected gene have been identified in only a very small number of patients. |
| **Inheritance** | Autosomal dominant. |
| **Chromosomal location** | 3p25 |
| **Genes** | *PPARG* (peroxisome proliferator-activated receptor γ). |
| **Mutational spectrum** | Two missense mutations have been identified: Pro167Leu and Val290Met. |
| **Effect of mutation** | Decrease PPAR-γ function, a key regulator of adipocyte differentiation. |

**Diagnosis**   The diagnosis must be based on finding a mutation in the *PPARG* gene, since the clinical syndrome is not specific. Mutations should be sought in patients with severe insulin resistance, acanthosis nigricans, and hyperandrogenism.

**Counseling issues**   *De novo* mutations have been reported. In these cases, the patient's siblings and offspring are not at significant risk. However, the children of affected individuals have a 50% risk of inheriting the disease.

**Notes**   A new class of drugs, the thiazolidinediones, has recently been approved for use in type 2 diabetes mellitus patients with insulin resistance. These drugs are specific ligands for PPAR-γ, activating the transcription factor and thus increasing tissue insulin sensitivity. Since activation of *PPARG* can increase insulin sensitivity, it was reasoned that mutations in PPAR-γ may cause insulin resistance. This led to mutation analysis in a cohort of severely resistant type 2 diabetic patients, and the discovery of two affected individuals. A direct β-cell function for *PPARG* has been suggested.

**References**   Barroso I, Gurnell M, Crowley VE et al. Dominant negative mutations in human PPARgamma associated with severe insulin resistance, diabetes mellitus and hypertension. *Nature* 1999;402:880–3.

# Thiamine-responsive Megaloblastic Anemia Syndrome

(also known as: TRMA; Roger's syndrome; thiamin-responsive anemia syndrome)

| | |
|---|---|
| **MIM** | 249270 |
| **Clinical features** | The syndrome complex of TRMA includes megaloblastic anemia, diabetes, and sensorineural deafness. Amenorrhea has been reported, but it is not clear if this is part of the syndrome or an incidental finding. One patient was reported with a cerebrovascular accident at the age of 18 years. The connection between this and the metabolic defect is not clear. |
| **Primary tissue or gland affected** | β cells. |
| **Other organs, tissues, or glands affected** | Ears, bone marrow. |
| **Age of onset** | The initial findings typically develop during the first year of life, though diabetes may develop later. |
| **Epidemiology** | Only a few cases have been described. Mutation analysis in selected target populations has not been reported. |
| **Inheritance** | Autosomal recessive. |
| **Chromosomal location** | 1q23.3 |
| **Genes** | SLC19A2 (thiamine transporter protein). |
| **Mutational spectrum** | Missense, nonsense, and deletion mutations have been described. One or two base deletions in the coding sequence result in a phase shift and premature termination of translation. |

**Effect of mutation**    Mutations result in the production of a defective thiamine transporter protein.

**Diagnosis**    The combination of diabetes with sensorineural deafness suggests a mitochondrial disorder; however, the finding of megaloblastic anemia is specific for this syndrome. Definitive diagnosis is based on mutation analysis of the *SLC19A2* gene.

**Counseling issues**    Although high-dose thiamine treatment corrects the anemia and may also have a beneficial effect on the diabetes, it is not known if very early treatment can modify the onset of hearing loss.

**Notes**    This gene falls within a region identified during a whole-genome search for diabetes-susceptibility genes in Pima Indians. However, mutation analysis failed to identify any variants in this population.

**References**    Neufeld EJ, Fleming JC, Tartaglini E et al. Thiamine-responsive megaloblastic anemia syndrome: A disorder of high-affinity thiamine transport. *Blood Cells Mol Dis* 2001;27:135–8 (Review).

Thameem F, Wolford JK, Bogardus C et al. Analysis of slc19a2, on 1q23.3 encoding a thiamine transporter as a candidate gene for type 2 diabetes mellitus in Pima Indians. *Mol Genet Metab* 2001;72:360–3.

# Insulin Receptor Defects

(also known as: leprechaunism; Donohue syndrome; Rabson–Mendenhall syndrome; type A insulin resistance)

| | |
|---|---|
| **MIM** | 246200, 262190, 147670 |
| **Clinical features** | Mutations in the insulin receptor can cause insulin-resistance syndromes that range from mild to severe. Severe insulin resistance can cause either leprechaunism (Donohue syndrome) or Rabson–Mendenhall syndrome, two related but distinct clinical syndromes. Both are associated with dysmorphic features, intrauterine and postnatal growth retardation, lack of subcutaneous fat, acanthosis nigricans, enlargement of genitalia, hirsutism, fasting hypoglycemia, and postprandial hyperglycemia. Children with leprechaunism usually die within the first year of life. |

Rabson–Mendenhall syndrome includes premature and dysplastic dentition, coarse facial features, and pineal hyperplasia. Children with this syndrome are born with fasting hypoglycemia, but typically develop severe diabetes with ketoacidosis later in childhood. The etiology of the hypoglycemia is not clear. Insulin levels are extremely high, and glucose levels fail to respond to exogenous insulin.

Other insulin receptor mutations are associated with insulin resistance syndrome type A, which is characterized by markedly elevated insulin levels, signs of virilization, and polycystic ovaries. All of the reported patients are female, and most are black. However, mutations in the insulin receptor (*INSR*) gene have been identified in only a minority of patients with this syndrome, and the molecular etiology is unknown for most patients. Recently, the Val985met *INSR* variant was found at increased frequency (4.4% vs. 1.8%) in diabetic patients when compared with ethnically matched nondiabetic controls. This finding suggests that subtle receptor variants may play a role in determining the risk of developing type 2 diabetes mellitus.

| Primary tissue or gland affected | Fat, muscle, liver. |
|---|---|
| Age of onset | Birth. |
| Epidemiology | Severe insulin receptor defects are exceedingly rare, and have been reported in various ethnic groups. |
| Inheritance | Autosomal recessive. |
| Chromosomal location | 19p13.2 |
| Genes | *INSR* (insulin receptor). |
| Mutational spectrum | Missense and termination mutations have been reported. |
| Effect of mutation | Mutations result in varying degrees of receptor dysfunction, and can be divided into several classes according to how the mutation modifies receptor function: class 1, impaired receptor biosynthesis; class 2, impaired transport of the receptor to the cell surface; class 3, decreased affinity for insulin binding; class 4, impaired tyrosine kinase activity; and class 5, accelerated receptor degradation. |
| Diagnosis | Diagnosis of the severe insulin-resistance syndromes is based on finding the typical clinical characteristics along with extremely high insulin levels – as much as 1000 times the upper limit of normal. Definitive diagnosis is confirmed by genetic analysis of the *INSR* gene. Type A insulin-resistance syndrome is diagnosed clinically by demonstrating markedly elevated fasting insulin levels in the presence of acanthosis nigricans and virilization. *ISNR* gene-coding sequence mutations are found in only a minority of patients. |
| Counseling issues | The overall prognosis for the severe insulin-resistance syndromes (leprechaunism and Rabson–Mendenhall syndrome) is very poor, and is not significantly improved by currently available treatment. |

**Notes**

The overall importance of insulin receptor variants as risk factors for type 2 diabetes mellitus is not known, but is thought to be quite limited. Interestingly, two insulin receptor mRNA transcripts resulting from alternative splicing of exon 11 have been described. Tissue distribution of the two isoforms is tightly controlled, and each has somewhat different functional characteristics. Abnormal relative distribution of these splice variants could result in subtle differences in insulin sensitivity. This could be caused by mutations outside the coding region of this gene, or genetic variants in components of the mRNA processing apparatus.

**References**

Longo N, Wang Y, Pasquali M. Progressive decline in insulin levels in Rabson–Mendenhall syndrome. *J Clin Endocrinol Metab* 1999;84:2623–9.

# Insulin Gene Mutations

(also known as: familial hyperproinsulinemia)

| | |
|---|---|
| **MIM** | 176730 |
| **Clinical features** | Patients are typically thin and show no evidence of resistance to exogenous insulin. Diabetes may appear in childhood or early adulthood, and thus may be phenotypically similar to maturity-onset diabetes of the young (see page 183). |
| **Primary tissue or gland affected** | Pancreatic β cells. |
| **Age of onset** | Diabetes, if present, is usually diagnosed during late childhood or during adulthood. Hyperproinsulinemia is present at birth, but will only be diagnosed when insulin levels are measured during family screening, or for another, usually unrelated reason. |
| **Epidemiology** | Families with these mutations are rare and appear to be more common in the Japanese population. |
| **Inheritance** | Autosomal dominant. |
| **Chromosomal location** | 11p15.5 |
| **Genes** | *INS* (insulin). |
| **Mutational spectrum** | Four mutations have been identified within the coding sequence of the insulin B chain: Phe25Leu, Phe24Ser, Val3Leu, and Asp10His. These appear to be assoicated with type 2 diabetes. Three mutations have been identified at codon 65, altering Arg to Leu, His, or Pro. Arg65Leu is associated with diabetes, as is Arg65His in the Japanese population. However, the same mutation in a white family was not associated with diabetes, and the Arg65Pro mutation was similarly not associated with glucose intolerance. |

**Effect of mutation**   The Phe25Leu, Phe24Ser, and Val3Leu mutations alter the binding of insulin to its receptor. The Asp10His mutation results in abnormal insulin trafficking and secretion as unprocessed proinsulin through the constitutive pathway. The other mutations alter the cleavage site, and thus prevent the normal processing of proinsulin to insulin.

**Diagnosis**   The diagnosis is usually suspected when insulin levels are measured for some unrelated reason. Several of the affected families were identified by chance during clinical studies that included the measurement of insulin levels in diabetics and controls.

**Counseling issues**   As stated above, some of the mutations have been associated with type 2 diabetes mellitus, and thus patients with these mutations should be urged to modify their lifestyle to decrease the risk of hyperglycemia. Diabetes responds well to exogenous insulin, since, as opposed to patients with insulin receptor mutations, there is no resistance to structurally normal insulin.

**References**   Roder ME, Vissing H, Nauck MA. Hyperproinsulinemia in a three-generation Caucasian family due to mutant proinsulin (Arg65-His) not associated with impaired glucose tolerance: The contribution of mutant proinsulin to insulin bioactivity. *J Clin Endocrinol Metab* 1996;81:1634–40 (Review).

Warren-Perry MG, Manley SE, Ostrega D et al. A novel point mutation in the insulin gene giving rise to hyperproinsulinemia. *J Clin Endocrinol Metab* 1997;82:1629–31.

# Transient Neonatal Diabetes Mellitus

(also known as: TNDM; diabetes mellitus, transient neonatal [DMTN])

| | |
|---|---|
| **MIM** | 601410 |
| **Clinical features** | Patients are born with intrauterine growth retardation (IUGR), which is possibly an indication of hypoinsulinemia *in utero*. Hyperglycemia occurs shortly after birth, and insulin is required to maintain euglycemia. Some infants have macroglossia, which appears to be associated with paternal isodisomy or duplication. In about half of infants the diabetes is transient and resolves within the first 6 months of life (median: 3 months). The incidence of type 2 diabetes in later life appears to be elevated, suggesting a long-lasting β-cell defect. |
| **Primary tissue or gland affected** | Pancreatic β cells. |
| **Age of onset** | Hyperglycemia is typically evident in the first 6 weeks of life. |
| **Epidemiology** | Estimated to occur in 1:400 000 neonates. |
| **Inheritance** | Imprinting. |
| **Chromosomal location** | 6q24 |
| **Genes** | Unknown. |
| **Mutational spectrum** | Chomosomal duplication of 6q24 or paternal isodisomy of the same chromosomal region (UPD6) is found in some, but not all, patients. Some patients may have abnormal DNA methylation patterns on Ch6p24, suggesting abnormal imprinting. |
| **Effect of mutation** | The genetic defect (UPD6) may relate to an imprinted gene located on chromosome 6q24. The hypothesis is that duplication or paternal isodisomy causes bi-allelic expression of a gene that is normally expressed from a single allele. |

**Diagnosis**

The diagnosis is made by identifying hyperglycemia in an IUGR infant. The transient nature of the disease can only be determined by re-evaluation at age 6 months, a point at which transient disease would be expected to have resolved. In some patients, the hyperglycemia does not resolve and diabetes is persistent throughout life.

**Counseling issues**

The finding of uniparental disomy in this region of chromosome 6 suggests that the disease will be transient, and that the chance of recurrence in subsequent pregnancies is extremely low.

**Notes**

A significant percentage of patients will develop classical type 2 diabetes later in life. Recent studies suggest that the ZAC/PLAGL1 transcriptional regulator may be involved in the pathogenesis of this syndrome.

**References**

Arthur EI, Zlotogora J, Lerer I et al. Transient neonatal diabetes mellitus in a child with invdup(6)(q22q23) of paternal origin. *Eur J Hum Genet* 1997;5:417–9.

Gardner RJ, Mackay DJ, Mungall AJ et al. An imprinted locus associated with transient neonatal diabetes mellitus. *Hum Mol Genet* 2000;9:589–96.

# Pancreatic β-cell Agenesis with Neonatal Diabetes Mellitus

**MIM**                  600089

**Clinical features**    Patients have permanent, insulin-dependent diabetes from birth. This disease can be differentiated clinically from pancreatic agenesis (MIM 260370) since exocrine function is normal.

**Primary tissue or gland affected**    Pancreatic β cells.

**Age of onset**         Birth.

**Epidemiology**         Extremely rare, with only a few well-documented cases reported.

**Inheritance**          Autosomal recessive.

**Chromosomal location**    6

**Genes**                Unknown.

**Mutational spectrum**    Paternal uniparental isodisomy of a portion of chromosome 6 has been reported in a patient who also had methylmalonic acidemia (MIM 251000).

**Effect of mutation**    Since only isolated cases have been reported, the hypothesis is that an imprinted gene on chromosome 6 is either not expressed or is overexpressed.

**Diagnosis**            The diagnosis is suspected when neonatal diabetes is seen with a structurally normal pancreas.

**Notes**                This may be a variant of transient neonatal diabetes mellitus (MIM 601410; see previous entry).

**References**

Abramowicz MJ, Andrien M, Dupont E et al. Isodisomy of chromosome 6 in a newborn with methylmalonic acidemia and agenesis of pancreatic beta cells causing diabetes mellitus. *J Clin Invest* 1994;94:418–21.

# Pancreatic Agenesis

(also known as: pancreatic hypoplasia, congenital)

| | |
|---|---|
| MIM | 260370 |
| Clinical features | Patients are born with intrauterine growth retardation (IUGR), neonatal insulin-dependent diabetes, and pancreatic exocrine insufficiency. Heterozygotes for mutations in this gene cause one form of maturity-onset diabetes of the young (see MODY4, MIM 606392, page 183). |
| Primary tissue or gland affected | Pancreatic β cells. |
| Other organs, tissues, or glands affected | Pancreatic exocrine tissue. |
| Age of onset | Birth. |
| Epidemiology | Extremely rare; only a few cases have been reported. |
| Inheritance | Autosomal recessive. |
| Chromosomal location | 13q12.1 |
| Genes | *IPF1* (insulin promoter factor 1; also known as *PDX1* [pancreatic–duodenal homeobox 1], *STF1* [somatostatin transcription factor 1], *GSF1* [glucose-sensitive factor 1] and *IDX1* [islet–duodenal homeobox 1]). |
| Mutational spectrum | One mutation, a single base deletion resulting in a frameshift and premature termination (FS123Ter), has been published. Missense mutations that appear to be associated with an increased risk of type 2 diabetes mellitus (MIM 125853) have been reported. |

**Effect of mutation**   The first mutation described resulted in a truncated, nonfunctional protein. Presumably any mutation that results in failure to produce a functional transcription factor will result in this syndrome.

**Diagnosis**   Clinical diagnosis is based on finding absent or very low insulin and C-peptide levels, pancreatic exocrine insufficiency, and absent pancreas on pancreatic imaging. Definitive diagnosis requires identification of mutations in the *IPF1* gene. It is likely that mutations in other genes critical for the development of the pancreas will produce a similar clinical syndrome.

**Counseling issues**   Transmission of this disease is autosomal recessive. However, heterozygotes have much milder disease, characterized as mild type 2 diabetes, and appearing during early adulthood (see MODY4, MIM 606392, page 183).

**References**   Stoffers DA, Zinkin NT, Stanojevic V et al. Pancreatic agenesis attributable to a single nucleotide deletion in the human IPF1 gene coding sequence. *Nat Genet* 1997;15:106–10.

# Hyperinsulinism of Infancy

(also known as: HI; persistent hyperinsulinemic hypoglycemia of infancy [PHHI]; nesidioblastosis; HI-KATP)

| | |
|---|---|
| **MIM** | 601820, 256450 |

**Clinical features**

The primary clinical feature of HI is hypoglycemia, which may be difficult to diagnose in the neonate period since presenting symptoms may be nonspecific. Once hypoglycemia is identified, it must be treated aggressively to prevent irreversible brain damage. The disease is most frequently caused by a defect in the β-cell ATP-dependent potassium channel (KATP). This channel maintains the β-cell membrane in a hyperpolarized state, and couples the metabolic state of the cell to insulin secretion. When the channel is defective, the membrane is chronically depolarized, voltage-gated calcium channels are open, and the resulting influx of calcium causes unregulated insulin secretion. Other causes of hypoglycemia, such as defects in glucose production or counter-regulatory hormone deficiency, must be excluded. In some cases, insulin secretion can be suppressed with diazoxide, a KATP-channel opener, though most patients with severe, early-onset disease do not respond to this drug. Somatostatin analogs may be useful in partially suppressing insulin secretion, though this drug also suppresses counter-regulatory hormones such as growth hormone and glucagon. The calcium channel blocker nifedipine has been reported to be effective in a minority of cases. Near total pancreatectomy may be needed to prevent recurrent hypoglycemia, though in cases successfully treated without pancreatectomy, apparent clinical remission of hypoglycemia has been reported after months to years of treatment. Patients who undergo partial pancreatectomy appear to be at risk of developing diabetes during late childhood or early adolescence.

**Primary tissue or gland affected**

Pancreatic β cells.

| Age of onset | Typically during the first days of life. However, some patients develop hypoglycemia later in the neonatal period or even during early childhood. |
|---|---|
| Epidemiology | HI has been reported in most ethnic groups in the world. In outbred populations, the incidence is approximately 1:50 000; however, in some inbred populations an incidence of up to 1:2500 has been reported. There is an apparent increased incidence of disease in the Ashkenazi Jewish population, which can be traced to two founder mutations. |
| Inheritance | Autosomal recessive, autosomal dominant, other. |
| Chromosomal location | 11p15.1 |
| Genes | ABCC8 (sulfonylurea receptor, SUR1), CDKN1C (inward rectifying potassium channel, Kir6.2). |
| Mutational spectrum | Over 50 mutations have been reported, and include missense, nonsense, and splice-site mutations, as well as insertions and deletions. Most mutations are recessive, though dominant mutations have been recently described. Approximately one third of patients have focal disease caused by a paternally inherited mutation combined with a somatic loss of heterozygosity in a β-cell precursor. This results in clonal proliferation of defective β cells that secrete high levels of insulin in an unregulated manner. |
| Effect of mutation | Mutations result in decreased net channel activity. Some mutations appear to prevent the production of any SUR1 or Kir6.2 protein. Others result in defective channels that are either not transported to the membrane (trafficking defects), or are present on the membrane, but fail to respond appropriately to ATP and/or ADP regulation. |
| Diagnosis | Hyperinsulinism may be suspected when the glucose requirement to prevent hypoglycemia is greater than the normal glucose |

requirement for the age of the patient (8 mg/[kg × min] for the neonate). This suggests increased glucose utilization, as opposed to inadequate production, which can be seen in a wide variety of metabolic and endocrine disorders. Unless a highly sensitive insulin assay is used, it may be difficult to document inappropriately elevated insulin levels. Surrogate measurements of insulin action, including suppressed ketone body production, increased glycemic response to glucagon, and glucose requirement to prevent hypoglycemia, may be particularly useful in making the diagnosis.

**Counseling issues**

Identification of the responsible mutation is necessary to accurately define the genetic risk to siblings and other family members, since recessive, dominant, and focal disease cannot be reliably identified on clinical findings alone. If focal HI is demonstrated histologically, the genetic risk to subsequent siblings is very low (probably <1%–2%). It is important to exclude other forms of HI, such as hyperinsulinism–hyperammonemia syndrome (HI/HA, MIM 138130; see next entry), since HI/HA is transmitted as an autosomal dominant trait, and probands frequently represent new mutations. In the absence of an accurate genetic diagnosis, families must be made aware that the risk of disease in subsequent children may be as high as 50% or as low as 1%–2%. Counseling is further complicated by the finding that as many as 40% of cases may be caused by mutations in other, as yet unidentified, genes.

**Notes**

Although most mutations are recessive, some dominant mutations have been described. As many as one third of cases may have a focal variant of the disease. These patients inherit a single mutant allele from their fathers. During pancreatic development, a single precursor cell loses the maternal allele of chromosome 11p15-ter. The progeny of this cell contain only mutant *SUR1* genes, and lack imprinted growth-suppressing genes on the lost maternal allele (loss of heterozygosity; LOH). The result is a proliferating, focal lesion composed of β cells that secrete insulin in an unregulated manner (focal HI). If this region can be identified and selectively resected, a complete cure can be attained. The incidence of LOH during pancreas development is not known, but has been estimated at ≤1%–2%.

**References**

Glaser B, Landau H, Permutt MA. Neonatal hyperinsulinism. *Trends Endocrinol Metabol* 1999;10:55–61.

Glaser B, Thornton P, Otonkoski T et al. The genetics of hyperinsulinism. *Arch Dis Child Fetal Neonatal Ed* 2000;82:F79–F86 (Review).

# Hyperinsulinism–Hyperammonemia Syndrome

**MIM**  606762

**Clinical features**  Unlike most patients with hyperinsulinism of infancy (HI) due to ATP-dependent potassium channel mutations (see the previous entry, MIM 601820), birth weights are normal for gestational age. Hypoglycemia is relatively mild, and may be exacerbated by a high protein diet. These patients may have markedly exaggerated insulin responses to leucine stimulation – many patients previously diagnosed with 'leucine-sensitive hypoglycemia' may have actually had this genetic defect. Approximately 40% of patients can be treated with modification of diet alone. Most of the remaining 60% will respond well to diazoxide treatment. Rarely, partial pancreatectomy is needed to control the hyperinsulinism. The hyperammonemia is constant, unaffected by diet or glycemic control, and appears to be asymptomatic. As many as 50% of cases have developmental delays or mental retardation, thought to be caused by unrecognized recurrent hypoglycemia.

**Primary tissue or gland affected**  Pancreatic $\beta$ cells.

**Age of onset**  Mild symptoms may be present at birth, though the diagnosis is frequently missed. The median age at diagnosis is 11 months.

**Epidemiology**  The incidence of this syndrome is not known; however, it is thought that only a small percentage (approximately 5%) of patients with HI (incidence 1:50 000) have disease caused by mutations in the *GLUD1* gene. As discussed below, many cases are caused by *de novo* mutations and thus the disease is found in all ethnic groups.

**Inheritance**  Autosomal dominant.

**Chromosomal location**  10q23.3

| | |
|---|---|
| **Genes** | *GLUD1* (glutamate dehydrogenase). |
| **Mutational spectrum** | All the mutations identified so far are missense mutations, either in the regulatory domain coded by exons 11 and 12, or in the GTP-binding domain coded by exons 6 and 7. |
| **Effect of mutation** | Gain of function, since the mutations reduce the sensitivity of allosteric inhibition by GTP. |
| **Diagnosis** | The clinical diagnosis is based on finding hyperinsulinemic hypoglycemia in the presence of elevated ammonia levels. Absolute levels of ammonia may vary from slightly above the reference range to up to 4 times the upper limit of normal. The presence of this specific syndrome can be confirmed by genetic analysis of the *GLUD1* gene. However, in a recent cohort of 65 patients with the hyperinsulinemic–hyperammonemia clinical syndrome, *GLUD1* mutations were identified in only 50 patients. The fact that no mutations were identified in the other 15 patients suggests that an identical clinical syndrome can also be caused by mutations in other genes. |
| **Counseling issues** | The disease is transmitted as an autosomal dominant trait, though a large percentage of cases are caused by *de novo* mutations. In a recent study, 15 out of 19 cases were found to be caused by *de novo* mutations. Therefore, identification of the mutation and parental screening are necessary before the risk to subsequent children can be assessed. If mutation analysis is not possible or is negative, a detailed metabolic study may help to identify mild disease in one of the parents. |
| **Notes** | See also: hyperinsulinism of infancy, MIMs 256450, 601820, 138130, and 602485. |
| **References** | Stanley CA, Lieu YK, Hsu BY et al. Hyperinsulinism and hyperammonemia in infants with regulatory mutations of the glutamate dehydrogenase gene. *N Engl J Med* 1998;338:1352–7. |

# Hyperinsulinism–Glucokinase

(also known as: HI–GCK; persistent hyperinsulinemic hypoglycemia of infancy–glucokinase)

| | |
|---|---|
| **MIM** | 602485 |
| **Clinical features** | Patients with this syndrome have both fasting and postprandial hypoglycemia. Prolonged fasting results in stable hypoglycemia of approximately 2 mmol/L. Insulin-induced hypoglycemia below this threshold results in the suppression of endogenous insulin secretion. Insulin secretion is stimulated by glucose or mixed meal. Postprandial hypoglycemia is caused by a delayed turn-off in insulin secretion. Thus, the clinical syndrome is caused by an abnormal set-point at which glucose-stimulated insulin secretion is initiated. The published mutations result in a moderate decrease in this set-point, and this is a relatively mild phenotype that responds well to diazoxide therapy. Mutations that produce more marked changes in glucokinase activity would be expected to produce far more profound hypoglycemia. |
| **Primary tissue or gland affected** | Pancreatic β cells. |
| **Age of onset** | Although hypoglycemia is present from childhood, it can be mild, and clinical diagnosis may be delayed until later in life. |
| **Epidemiology** | Two familes with this syndrome have been reported to date, but the actual prevalence among HI patients is unknown. |
| **Inheritance** | Autosomal dominant. |
| **Chromosomal location** | 7p15-p13 |
| **Genes** | *GCK* (glucokinase). |
| **Mutational spectrum** | Two missense mutations have been described (V455M and A456V). |

**Effect of mutation**   The mutation results in increased affinity of the enzyme for glucose. Thus, enzyme activity is increased at low glucose levels.

**Diagnosis**   The clinical diagnosis can be made by demonstrating insulin-secretory dynamics consistent with a lowered glucose threshold, but a definitive diagnosis requires demonstration of a gene mutation that causes the appropriate changes in enzyme function. Patients respond well to diazoxide treatment.

**Counseling issues**   The syndrome is transmitted as an autosomal dominant trait, and, unlike other forms of hyperinsulinemic hypoglycemia, published cases have been easily and completely controlled by medication and diet.

**Notes**   In the same gene, mutations that decrease the enzyme's affinity for glucose cause one form of maturity-onset diabetes of the young (MODY2, MIM 125851).

**References**   Glaser B, Kesavan P, Heyman M et al. Familial hyperinsulinism caused by an activating glucokinase mutation. *N Engl J Med* 1998;338:226–30.

# Beckwith–Wiedemann Syndrome

(also known as: BWS; exomphalos–macroglossia–gigantism syndrome [EMG])

| | |
|---|---|
| **MIM** | 130650 |

**Clinical features**
There are a large number of clinical findings, some of which are quite nonspecific. Typical findings include generalized overgrowth or hemihypertrophy, linear ear lobe creases, and omphalocele. About 30% of patients will have hyperinsulinemic hypoglycemia, apparently due to pancreatic β-cell hyperplasia. Most of these have transient disease, treatable with glucose and diazoxide; however, some have severe disease that may require partial pancreatectomy. Other clinical features include a large fontanel, prominent occiput, coarse facial features, renal medullary dysplasia, and adrenocortical cytomegaly. There is an increased incidence of Wilms' tumor, hepatoblastoma, adrenal carcinoma, and gonadoblastoma.

**Primary tissue or gland affected**
Pancreatic β cells.

**Age of onset**
Diagnosis can usually be made at birth.

**Epidemiology**
The incidence has been estimated at 1:13 700 births.

**Inheritance**
Autosomal dominant.

**Chromosomal location**
11p15.5

**Genes**
Abnormalities in the expression of imprinted genes in the region of Ch11p15.5, CDKN1C (also known as *p57[KIP2]*), *IGF2* (insulin-like growth factor 2), *H19*.

**Mutational spectrum**
The genetic findings for this syndrome are complex and variable. In 20% of patients, paternal uniparental disomy for 11p15.5 can be found in the absence of cytogenetic abnormalities. Others display

maternal imprinting center relaxation of *IGF2* and/or *H19*. In some, abnormal methylation patterns can be found. In 2%–3% of cases there are cytogenetic abnormalities due to either paternal duplications, or maternal translocations or inversions. All of these have the common end result of abnormal expression patterns of imprinted genes located on chromosome 11p15.5. Recently, inactivating point mutations in the cell-cycle regulating gene *p57(KIP2)* have been reported in a few patients. This gene is expressed and imprinted in pancreatic β cells.

**Effect of mutation**  Disruption of the normal imprinting of genes in this region appears to play an important role in the development of BWS.

**Diagnosis**  The diagnosis is based on the clinical syndrome. Identification of any of the above-listed genetic abnormalities will confirm the diagnosis. However, no genetic abnormality can be identified in a significant proportion of patients.

**Counseling issues**  The genetics of BWS are complex, and autosomal dominant inheritance cannot fully explain the disease. Several sets of monozygotic twins discordant for the disease have been reported.

**References**  DeBaun MR, Niemitz EL, McNeil DE et al. Epigenetic alterations of H19 and LIT1 distinguish patients with Beckwith–Wiedemann syndrome with cancer and birth defects. *Am J Hum Genet* 2002;70:604–11.

# Disorders of Gluconeogenesis and Glycogenolysis

(major specific syndromes: glucogen storage disease; hereditary fructose intolerance; galactosemia; maple syrup urine disease [MSUD])

| | |
|---|---|
| **MIM** | 240600, 232200, 229600, 248600, 251000 |
| **Clinical features** | Although not strictly endocrine diseases, this class of disorders is briefly described since patients may present with hypoglycemia, and initially be seen by an endocrinologist. Since glucose is the obligate energy source of the central nervous system, multiple mechanisms have developed to prevent hypoglycemia. During a short fast, glycogenolysis is the primary mechanism by which hypoglycemia is prevented. Any defect that prevents the production of glycogen during periods of energy excess, or the breakdown of glycogen when needed, will result in fasting hypoglycemia. Gluconeogenesis from amino acid substrates becomes the primary source of glucose after the first 12–24 hours of fasting. The severity of the hypoglycemia and associated defects, such as liver disease, encephalopathy, and acidosis, will depend on the precise nature of the defect. **Figure 4** demonstrates the major pathways that may be involved. |

Glucose-6-phosphatase (G-6-Pase) deficiency can be caused by mutations in the enzyme itself, or in any of the three translocase systems that allow the entry of glucose-6-phosphate into the endoplasmic reticulum. Since this enzyme is required for both gluconeogenesis and glycogenolysis, hypoglycemia occurs after short fasts and is severe. Glycogen synthase deficiency results in both postprandial hyperglycemia and fasting hypoglycemia, since glycogen synthesis is an important pathway in the disposal of excess glucose after a meal, and glycogen is a critical source of glucose during a short fast. Since the gluconeogenic pathway is intact, hypoglycemia may be less severe than with G-6-Pase mutations.

Disorders of glyconeogenesis, such as fructose-1,6-diphosphatase deficiency, will result in hypoglycemia only after a longer fast or

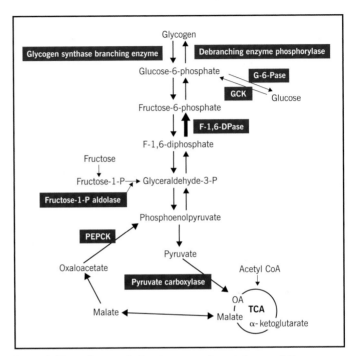

**Figure 4.** The major steps in the gluconeogenic and glycogenolytic pathways. Mutations in the enzymes highlighted may result in hypoglycemia. CoA: coenzyme A: F-1,6-DPase: fructose-1,6-diphosphatase; GCK: glucokinase; G-6-Pase: glucose-6-phosphatase; OA: oxaloacetate; PEPCK: phosphoenolpyruvate carboxykinase; TCA: tricarboxylic acid cycle.

during increased glucose demand, such as intercurrent illness. Other enzyme deficiencies, such as hereditary fructose intolerance, galactosemia, MSUD, propionic acidemia, and methylomalonic acidemia, can cause hypoglycemia by interfering with gluconeogenesis. For most of these, hypoglycemia is not the dominant feature, and is present only during stress situations such as prolonged fasting and intercurrent illness. Clinically, the hypoglycemia associated with all of these disorders is associated with suppressed insulin secretion, and can be corrected by supplying sufficient glucose to cover the basal glucose needs. This is in sharp contrast to insulin-induced hypoglycemia, where, in addition to the suppression of glucose production, there is a marked increase in glucose utilization, and very high rates of glucose infusion may be needed to prevent hypoglycemia.

| | |
|---|---|
| **Primary tissue or gland affected** | Liver. |
| **Age of onset** | Neonates. |
| **Epidemiology** | These are rare diseases, the incidence of which varies from 1:50 000 for methylmalonic acidemia, to isolated cases for glycogen synthase deficiency. |
| **Inheritance** | Autosomal recessive. |
| **Chromosomal location** | 17q21 (*G6PT*), 12p12.c (*GYS2*), 9q22.2-q22.3 (*FBP1*), 6p21 (*MCM*). |
| **Genes** | *G6PT* (glucose-6-phosphatase), *GYS2* (glycogen synthase), *FBP1* (fructose-1,6-diphosphatase deficiency), *MCM* (methylmalonyl CoA mutase). |
| **Mutational spectrum** | Mutations of different types have been reported in each of the genes responsible for the various specific disorders of gluconeogenesis and glycogenolysis. |
| **Effect of mutation** | Mutations cause defective enzyme function. The precise mechanisms vary for each enzyme involved. |
| **Diagnosis** | The precise diagnosis can be made by measuring various metabolites in the blood and specific enzyme activity in liver tissue. Some defects in glucose production can be detected by performing a glucagon stimulation test during hypoglycemia. In patients with insulin-induced hypoglycemia, the liver has ample glycogen stores, and high-dose glucagon (0.03 mg/kg up to 1 mg/kg intravenous) will result in an increase of >30 mg/dL in serum glucose levels. Patients |

with defective glycogenolysis or gluconeogenesis will have significantly lower responses.

**Counseling issues**  For adequate genetic counseling, precise metabolic and genetic diagnoses are essential.

**References**  Lteif AN, Schwenk WF. Hypoglycemia in infants and children. *Endocrinol Metab Clin North Am* 1999;28:619–46, vii (Review).

# Fatty-acid Oxidation Disorders

(major specific syndromes: carnitine deficiency; medium-chain acyl-CoA dehydrogenase deficiency [MCAD]; multiple acyl-CoA dehydrogenase deficiency [MADD]; glutaric aciduria type 2)

| | |
|---|---|
| **MIM** | 212140, 201450, 603584. 231680 |

**Clinical features**

Although not strictly endocrine diseases, patients with this class of disorders present with hypoglycemia, and thus may initially be seen by an endocrinologist. Fatty-acid $\beta$ oxidation is an important energy source, particularly during fasting and at times of metabolic stress. Disorders of fatty-acid oxidation can cause hypoglycemia, though the precise mechanism is not entirely known. One hypothesis is that failure to produce ketone bodies deprives tissues of an important energy source and makes them dependent on glucose as a primary energy source, even during fasting and metabolic stress. This increase in glucose utilization overtaxes the gluconeogenic machinery, and hypoglycemia results. Recently, hyperinsulinemic hypoglycemia has been reported in a few patients with short-chain 3-hydroxyacyl CoA dehydrogenase deficiency. The mechanism by which mutations in this enzyme cause hyperinsulinism is not known.

Although this class of disorders can be caused by defects in any of a large number of enzymes required for fatty-acid oxidation (see **Figure 5**), the clinical presentation is similar for all. Patients present with fasting hypoglycemia, low ketone body levels, and suppressed insulin levels. Metabolic derangement, typically associated with Reye's syndrome, such as encephalopathy, hyperammonemia, metabolic acidosis, and elevated transaminases, may be present. Myopathy and cardiomyopathy may also be prominent features. Acute symptoms are typically precipitated by fasting or acute intercurrent illness. Therefore, this class of disorders must be excluded before a patient suspected of having hypoglycemia is submitted to a diagnostic fast.

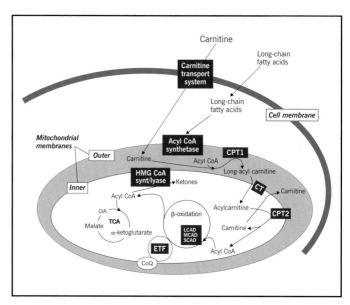

**Figure 5.** Fatty-acid oxidation. The major fatty acid oxidation pathways are shown, and the enzymes associated with defined genetic defects are highlighted in purple. CPT: carnitine palmitoyl transferase; CT: carnitine translocase; CoA: coenzyme A; CoQ: coenzyme Q; ETF: electron-transport flavoproteins; HMG: 3-hydroxy-3-methylglutaryl; LCAD, MCAD, SCAD: long-, medium-, and short-chain acyl coenzyme A dehydrogenases; OA: oxaloacetate; TCA: tricarboxylic acid cycle.

| | |
|---|---|
| **Primary tissue or gland affected** | Liver, muscle. |
| **Age of onset** | Onset is typically during infancy or early childhood. |
| **Epidemiology** | Rare. |
| **Inheritance** | Autosomal recessive. |
| **Chromosomal location** | 5q33.1 (OCTN2), 1p31 (MCAD), 15q23-q25 (MADD), 12q22-qter (SCAD), 11q13 (CPT1), 1p32 (CPT2). |

| Genes | OCTN2 (sodium ion-dependent carnitine transporter), MCAD (medium-chain acyl-CoA dehydrogenase), MADD (multiple acyl-CoA dehydrogenase deficiency), SCAD (short-chain acyl-CoA dehydrogenase), CPT1 (carnitine palmitoyltransferase 1), CPT2 (carnitine palmitoyltransferase 2). |
|---|---|
| Mutational spectrum | Mutations in a large number of different genes can cause this syndrome. Only a few are mentioned here. The interested reader is referred to a more detailed review of this topic (see reference below). |
| Effect of mutation | In general, mutations result in failure to produce a specific functioning enzyme, though different mechanisms may be present in different specific syndromes. |
| Diagnosis | A disorder in fatty-acid oxidation is suspected when a patient presents with hypoketotic hypoglycemia that is not associated with hyperinsulinism. The diagnosis can be confirmed by analyzing urine for organic acids, and measuring plasma carnitine and acylcarnitine levels. Electrospray tandem mass spectrometry can be used for screening and for rapid and precise diagnosis of most of these disorders. **Figure 5** shows the major pathways and enzymes required for fatty-acid oxidation. The highlighted enzymes are associated with different specific metabolic defects. |
| Counseling issues | Precise diagnosis is required for optimal counseling. |
| References | Lteif AN, Schwenk WF. Hypoglycemia in infants and children. *Endocrinol Metab Clin North Am* 1999;28:619–46. |

# Glycerol Kinase Deficiency

(also known as: GKD; hyperglycerolemia; GK1 deficiency)

| | |
|---|---|
| MIM | 307030 |

**Clinical features**

A variety of symptoms can be directly attributed to GKD, ranging from asymptomatic to episodes of severe metabolic decompensation with Reye's syndrome-like symptoms. These include vomiting, metabolic acidosis, ketotic hypoglycemia, and progressive lethargy leading to loss of consciousness. The clinical symptoms of GKD include episodic ketoacidosis, hypoglycemia, and/or apnea, usually associated with periods of metabolic stress, such as intercurrent illness or strenuous exercise. Some patients have psychomotor and/or mental retardation, but the association between these and the specific gene defect is unclear. When disease is caused by a deletion, other, adjacent genes can also be involved causing a contiguous gene syndrome (see 'Effect of mutation'). If the congenital adrenal hypoplasia gene (*NROB1*) is involved, patients typically present in the first months of life with failure to thrive, salt wasting, hypoglycemia, and hyperpigmentation. Hypogonadotropic hypogonadism is also frequently present. Deletion of the dystrophin gene (*DMD*) causes Duchenne muscular dystrophy, a progressive muscle disease that usually presents clinically at the time the infant starts to walk. Patients are usually wheelchair-bound by puberty. Another gene associated with mental retardation has been localized to this region, and the association between GKD and mental retardation may be another variant of the contiguous gene syndrome.

**Primary tissue or gland affected**

Liver.

**Age of onset**

Isolated GKD typically presents during early childhood (0–7 years of age). As many as two thirds of patients may be asymptomatic, though this may depend on the age of the patient, since symptoms can occur later in life. Diagnosis may be delayed until adulthood or until genetic screening is performed following the diagnosis of an index case in the family. Patients with contiguous gene syndrome, including adrenal hypoplasia and/or DMD, are typically diagnosed at an earlier age.

**Epidemiology**

More than 38 patients with isolated GKD and about 100 patients with contiguous gene syndromes, including GKD, have been reported. Given the marked variability in clinical findings, even in different patients from the same pedigrees, the actual incidence may be significantly higher.

**Inheritance**

X-linked.

**Chromosomal location**

Xp21.3-p21.2

**Genes**

*GK* (glycerol kinase).

**Mutational spectrum**

In patients with isolated GKD, one nonsense, seven missense, and two splice-site mutations have been described, in addition to one Alu insertion and two small deletions. GKD may also present as part of an X-linked contiguous gene syndrome, including deletions of the nuclear receptor gene (*NROB1*) causing congenital adrenal hypoplasia (MIM 300200), and the dystrophin gene causing DMD.

**Effect of mutation**

Point mutations and small deletions result in the lack of a functional enzyme. Larger deletions may result in a contiguous gene syndrome. The Xp21 locus includes at least six adjacent genes in the following orientation: *Xpter – NROB1 – GKD – DMD – CGD* (chronic granulomatous disease) *– OTCD* (ornithine transcarbamoylase deficiency) *– RP* (retinitis pigmentosa) *– Xcen*. Major deletions including more than one of these genes will cause a contiguous gene syndrome, the specific characteristics of which will be determined by the genes deleted. GKD is frequently found in combination with AHC and/or DMD. Association with other, adjacent diseases is much less common.

**Diagnosis**

GK catalyses the phosphorylation of glycerol to glycerol-3-phosphate, most of which then enters the Embden–Meyerhof pathway at the level of glyceraldehyde-3-phosphate. Enzyme defects result in hyperglycerolemia, and lack of availability of glycerol as a substrate for gluconeogenesis. Elevated plasma free glycerol levels will result in an overestimation of triglyceride levels if these measurements are performed by *in vitro* lipolysis followed by determination of glycerol levels. In the nonfasting state, glycerol contributes only a very small percentage to the total hepatic glucose output. However, during prolonged fasting, about 20% of gluconeogenesis is dependent on glycerol as a substrate. The biochemical diagnosis is suspected when hyperglycerolemia and glyceroluria are found in the presence of Reye's syndrome-like symptoms. Hyperketonemia is frequently present and may be the consequence of inadequate gluconeogenesis, though the possibility of underutilization of ketone bodies must also be considered. A definitive diagnosis relies on the identification of specific gene mutations. Genetic analysis is complicated by the fact that several regions of very high sequence homology have been identified, and include two pseudogenes and two expressed genes. Co-amplification of these genes in polymerase chain reaction-based mutation screening methods may hamper mutation analysis at the genomic level.

**Counseling issues**

Because of the severe, life-threatening nature of the disease, and the associated mental retardation, families of index cases frequently request genetic counseling. Once the genetic basis of disease in the index case has been established, carrier screening and prenatal diagnosis are possible. Because of the technical difficulties stemming from the presence of four genes with marked sequence similarity, biochemical testing may be used in place of, or in conjunction with, gene-based testing.

**References**

Sjarif DR, Ploos van Amstel JK, Duran M et al. Isolated and contiguous glycerol kinase gene disorders: A review. *J Inherit Metab Dis* 2000;23:529–47 (Review).

# 6. Genetic Defects of the Adrenal Cortex, Renal Electrolyte Balance, and Adrenal Medulla

# Defects in steroidogenesis in the adrenal cortex

The first eight entries in this chapter are caused by defects in genes that produce the enzymes necessary to convert cholesterol into the various adrenal steroid hormones. **Figure 1** shows the major steroid biosynthetic pathways and the critical enzymes involved. The reader should refer to this figure when reviewing the syndromes described in this section.

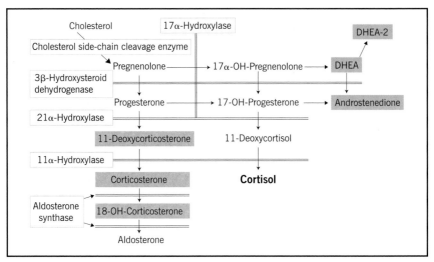

**Figure 1.** Adrenal steroidogenesis. The major enzymatic steps are shown. Steroids with mineralocorticoid activity are shown with a gray background. Androgenic steroids are shown with a purple background. Cortisol is the major glucocorticoid produced in man.

## Other defects in glucocorticoid production

Not all adrenal cortical defects are caused by mutations in the enzymes of the steroidogenic pathways. Primary glucocorticoid deficiency can be caused by a variety of other adrenal defects, including adrenocorticotropic hormone receptor defects, transcription factor mutations, and mutations in other proteins necessary for normal adrenal function. This is demonstrated in the next four syndromes presented.

# Abnormalities in renal electrolyte balance

Water and electrolyte imbalance can be caused by defects in mineralocorticoid production, or by defects at the level of the renal tubules. These syndromes are grouped together, since their clinical presentation may be similar. **Figure 2** shows the portion of the nephron affected by some of the primary renal defects described. **Figure 3** shows a schematic representation of the cortical collecting duct epithelial cell, highlighting the genetic defects that cause the syndromes described in this section. Although these are not hormonal in origin, the practicing endocrinologist is frequently called upon to evaluate patients with the mineral and electrolyte imbalances that are associated with these defects.

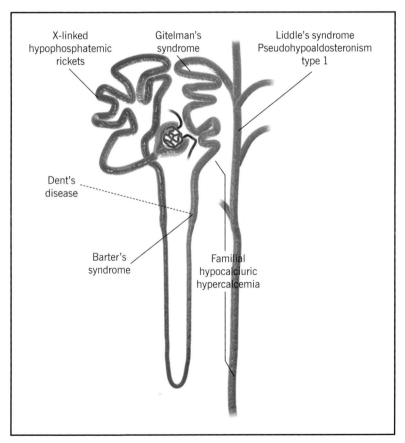

**Figure 2.** Location of genetic defects in renal mineral and electrolyte secretion. Reproduced with permission from Massachusetts Medical Society (*N Engl J Med* 1999;340:1177–87).

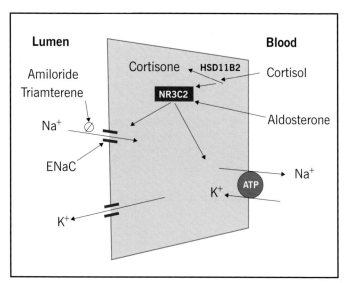

**Figure 3.** Schematic representation of the cortical collecting duct epithelial cell. Loss of function mutations in the aldosterone receptor (NR3C2) or the epithelial sodium channel (ENaC) will cause pseudohypoaldosteronism, characterized by urinary sodium loss and hyperkalemia. Activating mutations of ENaC will cause hypertension and hypokalemia (Liddle's syndrome, MIM 177200). Loss of function mutations in the *HSD11B2* gene result in pathologic stimulation of the aldosterone receptor by cortisol, resulting in apparent mineralocorticoid excess (MIM 218030)

# Adrenal tumors

Two familial syndromes of adrenal cortical tumors are discussed. Carney's syndrome is a complex disorder associated with benign or malignant adrenal tumors, and a wide variety of other endocrine and nonendocrine tumors. Hereditary adrenal cortical carcinoma is a very rare entity, the genetic etiology of which has not yet been completely established.

# Diseases of the adrenal medulla

Finally, three distinct syndromes whose primary endocrine pathology is familial pheochromocytoma are described. Familial pheochromocytoma is also an integral part of two other genetic syndromes, namely multiple endocrine neoplasia (MEN)2a (MIM 171400) and MEN2b (MIM 162300). Since these syndromes typically include other endocrinopathies, they are discussed in detail in Chapter 1.

In the von Hippel-Lindau, neurofibromatosis type 1, and MEN2 syndromes, pheochromocytoma may be the dominant presenting clinical feature, and the nature of the underlying syndrome may become evident only after more extensive evaluation. Alternatively, patients with these syndromes may present with signs or symptoms related to other aspects of the syndrome, and the pheochromocytoma may only be discovered after careful clinical evaluation. Similarly, the familial nature of the syndromes may only become evident after careful screening of family members.

Since undiagnosed pheochromocytoma can cause severe or even lethal complications, it is important that the clinician be alert to the existence of these syndromes and perform the necessary diagnostic tests as early as possible.

# Congenital Lipoid Adrenal Hyperplasia

**(also known as: adrenal hyperplasia I)**

| | |
|---|---|
| **MIM** | 201710 |
| **Clinical features** | Severe adrenal insufficiency appears shortly after birth. Mineralocorticoid and glucocorticoid deficiency cause hypotension, hyponatremia, hyperkalemia, vomiting, feeding problems, and hypoglycemia. Unless treatment with glucocorticoids and mineralocorticoids is initiated, affected infants die within days to weeks as a result of severe adrenal insufficiency. Because of the lack of testicular androgen production, male infants have female external genitalia. Female infants have normal genitalia at birth. Affected girls usually fail to undergo puberty, but some cases of spontaneous puberty have been reported. |
| **Primary tissue or gland affected** | Adrenal cortex. |
| **Other organs, tissues, or glands affected** | Testes and ovaries. |
| **Age of onset** | Prenatal. |
| **Epidemiology** | This condition is very rare, but occurs more frequently, often in a less severe form, in Korean and Japanese populations. |
| **Inheritance** | Autosomal recessive. |
| **Chromosomal location** | 8p11.2 |
| **Genes** | *StAR* (steroidogenic acute regulatory protein). |

| | |
|---|---|
| **Mutational spectrum** | At least 32 *StAR* mutations have been described, including nucleotide substitutions, deletions, insertions, splice mutations, and frame-shift mutations. |
| **Effect of mutation** | The StAR protein is an essential requirement in the transport of cholesterol from the outer to the inner mitochondrial membrane, the rate-limiting step in steroidogenesis. Mutant StAR proteins are nonfunctional. The inability of cholesterol to cross the mitochondrial membrane prevents acute-response steroidogenesis. Small amounts of steroids continue to be produced by a poorly understood, StAR-independent process. Prolonged accumulation of lipids within the cell explains the characteristic fatty appearance (lipoid hyperplasia), and results in cellular damage and cell death. Thus, in addition to the primary defect in steroidogenesis, a secondary ('second hit') effect leads to more severely compromised hormonal secretion. |
| **Diagnosis** | Low levels of hydrocortisone and aldosterone support the diagnosis of adrenal insufficiency. ACTH levels are markedly elevated. The differential diagnosis of lipoid adrenal hyperplasia includes 3β-hydroxysteroid dehydrogenase deficiency, which is characterized by elevated 17-hydroxypregnenolone levels. |
| **Counseling issues** | Although spontaneous puberty may occur in some female patients, menstrual cycles are anovulatory. Affected male and female patients are infertile. |
| **Notes** | The fetal ovary does not normally synthesize steroids during embryonic and early postnatal development. As a result, the 'second hit' mechanism of cell damage due to lipid accumulation does not occur in the ovary until puberty. The delayed onset of the 'second hit' may explain the occurrence of spontaneous puberty in some affected females. |
| **References** | Stocco DM. Clinical disorders associated with abnormal cholesterol transport: Mutations in the steroidogenic acute regulatory protein. *Mol Cell Endocrinol* 2002;191:19–25. |

# 3β-Hydroxysteroid Dehydrogenase Deficiency

(also known as: 3β-HSD; adrenal hyperplasia II; congenital adrenal hyperplasia [CAH] type II)

| | |
|---|---|
| MIM | 201810 |

**Clinical features**
The clinical presentation is variable and may include evidence of mineralocorticoid, cortisol, and sex steroid deficiency. Severe enzyme deficiency is associated with salt-losing crisis. Genetic males may present with male pseudohermaphroditism, and genetic females may be normal or mildly virilized at birth. Premature acne, sexual hair development, and accelerated growth may be seen. During adolescence, males may have spontaneous pubertal maturation, though gynecomastia and testicular hypertrophy may be seen. Genetic females may have evidence of androgen excess, primary or secondary amenorrhea, and polycystic ovaries. A 'nonclassical' or mild form of 3β-HSD has been described, and is characterized by hirsutism with or without premature pubarche. The diagnosis is based on elevated ACTH-stimulated δ5 steroids, such as 17-hydroxypregnenolone and dehydroepiandosterone. However, the validity of this biochemical diagnosis has been questioned, since no genetic mutations have been found in these patients.

**Primary tissue or gland affected**
Adrenal cortex.

**Other organs, tissues, or glands affected**
Gonads.

**Age of onset**
The severe form is diagnosed at birth, with salt-losing crisis and ambiguous genitalia. Milder forms can be diagnosed at any time during childhood, or at adolescence, at which time patients may present with varying degrees of hypogonadism.

**Epidemiology**
Among patients with 'classical CAH', the vast majority have 21-hydroxylase deficiency. Of the other well-described forms of CAH,

this is the least common. The incidence of the 'nonclassical' form is very controversial since diagnosis is based on unclear biochemical criteria and cannot be confirmed genetically.

| | |
|---|---|
| **Inheritance** | Autosomal recessive. |

**Chromosomal location**  1p13.1

**Genes**  *HSD3B2* (3β-hydroxysteroid dehydrogenase type 2).

**Mutational spectrum**  Missense, termination, and frame-shift mutations have been described.

**Effect of mutation**  The mutations result in decreased production of functional enzyme in the adrenal and gonads. In most cases, no functional enzyme is produced. Most patients wth nonsalt-wasting disease have missense mutations that may retain some enzymatic activity, thus explaining the milder phenotype.

**Diagnosis**  The clinical diagnosis of severe 3β-HSD deficiency is based on the finding of markedly elevated levels of 17-hydroxypregnenolone, low cortisol and aldosterone levels, and elevated ACTH levels. Sex hormone levels will also be decreased, depending on the age of the patient. Mutation analysis confirms the diagosis. 'Nonclassical', or mild, disease has been diagnosed on the basis of statistically elevated 17-hydroxypregnenolone:cortisol ratios. However, the lack of gene mutations in these patients suggests that this syndrome is caused by some other, as yet undefined, metabolic defect.

**Notes**  Two isoforms of the *HSD3B2* gene have been identified, located adjacent to each other on C1p11-13. The type 1 gene is primarily expressed in the placenta, mammary gland, and skin, whereas the type 2 gene is expressed in the adrenal and gonads. 3β-HSD-deficiency CAH is caused by mutations in the type 2 gene.

**References**

Pang, S. Congenital adrenal hyperplasia owing to 3β-hydroxysteroid dehydrogenase deficiency. *Endocrinol Metab Clin North Am* 2001;30:81–99, vi–vii (Review).

# Congenital Adrenal Hyperplasia: 17α-Hydroxylase Deficiency

(also known as: adrenal hyperplasia V)

| | |
|---|---|
| **MIM** | 202110 |
| **Clinical features** | Depending on the degree of impaired enzyme activity, affected male infants may present with ambiguous genitalia, or may appear to have a female phenotype consisting of female external genitalia, a blind vagina, no uterus or Fallopian tubes, and intra-abdominal testes. Patients fail to undergo spontaneous puberty. Elevated levels of 11-deoxysteroids result in hypertension and hypokalemia, though some patients may be normotensive at the time of the initial investigation. Patients with severe impairment of cortisol production may present with adrenal insufficiency. |
| **Primary tissue or gland affected** | Adrenal cortex. |
| **Age of onset** | Abnormal genital development in males occurs *in utero*. Affected males and females usually seek medical attention for the evaluation of delayed puberty. |
| **Epidemiology** | Approximately 120 cases of 17α-hydroxylase deficiency have been reported. Most patients have a combined defect in 17α-hydroxylase and 17,20-lyase activities, but some cases of isolated defects in only one enzyme have been described. |
| **Inheritance** | Autosomal recessive. |
| **Chromosomal location** | 10q24.3 |
| **Genes** | *CYP17* (encodes an enzyme which has 17α-hydroxylase activity as well as 17,20-lyase [cholesterol side-chain cleavage or desmolase] activity). |

**Mutational spectrum**   Missense, nonsense, deletion, and insertion mutations have been reported.

**Effect of mutation**   Except for premature stop codons, the precise mechanisms by which particular *CYP17* mutations result in loss of enzyme function are unknown. Decreased activity of 17α-hydroxylase results in decreased production of cortisol, which leads to increased ACTH-stimulated production of 11-deoxycorticosterone, corticosterone, and 18-hydroxydeoxycorticosterone. Reduced activity of 17,20-lyase in the adrenal gland and gonads results in decreased production of androgens and estrogens.

**Diagnosis**   Progesterone, 11-deoxycorticosterone, corticosterone, and 18-hydroxydeoxycorticosterone levels are elevated. Renin and aldosterone levels are low. Metabolites which require 17-hydroxylase or 17,20-lyase activity (cortisol, 11-deoxycortisol, 17-hydroxyprogesterone, dehydroepiandosterone, androstenedione, testosterone, and estradiol) are decreased. Luteinizing hormone and follicle-stimulating hormone concentrations are elevated.

**Counseling issues**   Most patients have been raised as females, including those with a 46,XY genotype. Alternatively, male patients could undergo extensive genital surgery and be treated with androgen replacement therapy. All patients raised as females need to receive estrogen replacement; progesterone should be given to those with a female genotype and uterus. Glucocorticoid administration decreases ACTH production and lowers the concentration of mineralocorticoid metabolites to the normal range.

**Notes**   The autoantigen associated with Addison's disease in type 1 polyendocrine autoimmunity syndrome (MIM 240300) is directed against 17α-hydroxylase. Promoter mutations in the *CYP17* gene may be associated with some forms of familial polycystic ovarian disease (MIM 184700) and premature male-patterned baldness (MIM 109200).

**References**

Yanase T. 17 alpha-Hydroxylase/17,20-lyase defects. *J Steroid Biochem Mol Biol* 1995;53:153–7 (Review).

# Congenital Adrenal Hyperplasia: 21-Hydroxylase Deficiency

(also known as: classical salt-wasting congenital adrenal hyperplasia [CAH]; classical simple virilizing CAH)

| | |
|---|---|
| **MIM** | 201910 |

| | |
|---|---|
| **Clinical features** | Female infants are born with ambiguous genitalia caused by intrauterine hyperandrogenism (see **Figure 4**). Both males and females present with Addisonian crisis, sometimes associated with hypoglycemia. The salt-wasting form is associated with hyponatremia and hyperkalemia with acidosis. A milder form of the disease, presenting only with hyperandrogenism in the female, is described separately. The incidence of benign adrenal tumors appears to be increased in patients with CAH and in mutation carriers. Malignant adrenal tumors have been rarely reported. In males, particularly with salt-wasting CAH, testicular adrenal rest tumors may occur and may suppress spermatogenesis. |
| **Primary tissue or gland affected** | Adrenal cortex. |
| **Age of onset** | Neonate to 4 years. The more severe, salt-wasting form is usually diagnosed in the immediate postnatal period with the onset of adrenal crisis. Females are typically diagnosed earlier than males because of ambiguous genitalia. |
| **Epidemiology** | Severe, salt-wasting disease is seen in approximately 1:20 000 live births, whereas the simple virilizing form has an incidence of about 1:60 000. A very high incidence of classical CAH (1:282–490) has been reported in the Yupik Eskimos of western Alaska, making prenatal diagnosis particularly important in this population. |
| **Inheritance** | Autosomal recessive. |

| Chromosomal location | 6p21.3 |
| --- | --- |
| Genes | *CYP21A2* (cytochrome P450, subfamily XX1A, polypeptide 2). About 90% of cases of CAH are caused by mutations in this gene. The remaining 10% are caused by mutations in other enzymes in the steroidogenesis pathways. |
| Mutational spectrum | More than 90% of mutations result from intergenic recombinations between *CYP21A2* and the closely linked pseudogene, *CYP21P*. At least 47 different mutations have been reported, and include missense mutations, insertions, deletions, splice-site mutations, and stop codons. |
| Effect of mutation | Salt-wasting disease is associated with a total lack of enzymatic activity, usually due to deletions, large conversions, or nonsense mutations. Simple virilizing disease is caused by severely reduced enzyme activity (about 1% of normal), and is usually associated with missense mutations. |
| Diagnosis | Diagnosis is based on finding low levels of cortisol, with very high levels of 17-hydroxyprogesterone and ACTH. Stimulation with exogenous ACTH is usually unnecessary, since basal levels of 17-hydroxyprogesterone are diagnostic. |
| Counseling issues | With early diagnosis, therapy is effective. In women at risk of bearing a homozygous female infant, dexamethasone treatment should be considered (see below). The risk of a patient with classical CAH having a partner who is an asymptomatic carrier of a severe CAH mutation is small, and depends on the population risk. |
| Notes | Treatment is based on glucocorticoid replacement and suppression of adrenal steroid production. The salt-wasting form requires the addition of fludrocortisone to prevent hyperkalemia and hypotension. Combination therapy with a glucocorticoid, antiandrogen, and aromatase inhibitor has been proposed. |

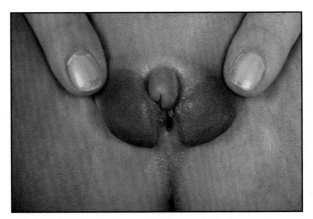

**Figure 4.** A genetic female with ambiguous genitalia caused by intrauterine hyperandrogenism.

Dexamethasone therapy during pregnancy may prevent the development of ambiguous genitalia in the homozygous female fetus; however, treatment must be initiated as soon as pregnancy is diagnosed, and continued until the sex and genotype of the fetus can be determined. Since only a homozygous female fetus will derive benefit from treatment, if all pregnant women at risk of having an affected fetus are treated, only one in eight will receive any benefit. Potential fetal risks of prenatal dexamethasone therapy include cardiac septal hypertrophy, hydrometrocolpos, and hydrocephalus. Although the precise risk of overt anomalies is not known, it appears to be low. The incidence of more subtle neurologic defects has not been adequately evaluated. Maternal complications, including cushingoid features, weight gain, hypertension, and hyperglycemia, may occur in as many as 10% of women. Therefore, treatment during pregnancy is still controversial.

**References**

White PC, Speiser PW. Congenital adrenal hyperplasia due to 21-hydroxylase deficiency. *Endocr Rev* 2000;21:245–91 (Review).

# Congenital Adrenal Hyperplasia: 11β-Hydroxylase Deficiency

(also known as: adrenal hyperplasia IV; CYP11B1 deficiency)

| | |
|---|---|
| **MIM** | 202010 |

**Clinical features**

Neonates may present with variable degrees of virilization, including an enlarged penis or clitoris, labial fusion, and a common urogenital sinus. Other presentations include precocious or early puberty in boys, and hirsutism and/or amenorrhea in adolescent girls. Androgen excess can lead to acne in adults. Hypertension occurs most commonly in patients with the more severe and early-onset forms of 11β-hydroxylase deficiency. Hypertension may be found in about two thirds of patients, and may or may not be accompanied by hypokalemic alkalosis. There is no correlation between hypertension, the presence of hypokalemia, or the degree of virilization. The degree of blood pressure elevation is weakly correlated with serum 11-deoxycorticosterone (DOC) levels. Surprisingly, some patients with documented 11β-hydroxylase deficiency have salt-wasting; the mechanism for this phenomenon and the heterogeneity in clinical features have not yet been clarified.

**Primary tissue or gland affected**

Adrenal cortex.

**Age of onset**

Virilization of affected females occurs *in utero*. The diagnosis may be recognized in neonates, children, or, in mild cases (late-onset), adults.

**Epidemiology**

11β-hydroxylase deficiency is the second most common type of congenital adrenal hyperplasia (CAH), representing 5%–8% of all CAH patients. The overall incidence is estimated to be 1:100 000 births. In Israel, the incidence among Jewish immigrants from Morocco is approximately 1:5000–7000 births.

| | |
|---|---|
| **Inheritance** | Autosomal recessive. |
| **Chromosomal location** | 8q21 |
| **Genes** | *CYP11B1* (encodes an isozyme of 11β-hydroxylase, located in the adrenal zona fasciculata, which is stimulated by ACTH [with cAMP as the second messenger], and whose main function is the conversion of 11-deoxycortisol [compound S] to cortisol and DOC to corticosterone). |
| **Mutational spectrum** | Approximately 31 mutations, including missense, nonsense, frame-shift, insertion, and deletion mutations, have been found. The most common mutation among Moroccan Israelis is a histidine for arginine substitution at codon 448. |
| **Effect of mutation** | The mutations which have been characterized to date result in near-total loss of function. Missense mutations cause loss of enzyme activity, while frame-shift and nonsense mutations block synthesis of the enzyme. Increased ACTH secretion, due to decreased synthesis of cortisol, stimulates the adrenal cortex to produce excessive amounts of 11-deoxycortisol, DOC, and adrenal androgens. |
| **Diagnosis** | The presence of hypertension, hypokalemia, and virilization in a neonate or child is typical of 11β-hydroxylase deficiency. Plasma renin activity is low, serum 11-deoxycortisol and DOC concentrations are elevated, and urinary tetrahydro metabolites of compound S and DOC are increased. One should be very cautious in interpreting steroid metabolite levels in the neonate period because of the cross-reactivity of various metabolites in most assays. In the milder, late-onset forms, basal levels of 11-deoxycortisol and DOC may be normal, and the diagnosis requires an ACTH stimulation test. Prenatal diagnosis has been performed by determining tetrahydro-11-deoxycortisol concentrations in amniotic fluid, and by sequencing the *CYP11B1* gene from chorionic villus biopsy samples. |

**Counseling issues**      Prenatal therapy with dexamethasone may be helpful in decreasing the severity of virilization in female infants with 11β-hydroxylase deficiency. Surgery may be required in infant girls with severe virilization of the genitalia. Glucocorticoid treatment is usually effective in treating hyperandrogenism and hypertension in children and adults, though additional antihypertensive medications may be needed. As with other types of CAH, growth and bone age need to be monitored carefully in children receiving glucocorticoid treatment.

**Notes**      In contrast to *CYP11B1*, the isozyme encoded by *CYP11B2* functions mainly to produce aldosterone in the zona glomerulosa. Mutations of *CYP11B2* result in salt-wasting due to aldosterone deficiency. Unequal crossing over of *CYP11B1* and the adjacent gene, *CYP11B2*, causes dexamethasone-suppressible hyperaldosteronism (MIM 103900).

**References**      White PC. Steroid 11 beta-hydroxylase deficiency and related disorders. *Endocrinol Metab Clin North Am* 2001;30:61–79, vi (Review).

# Aldosterone Deficiency

(also known as: isolated aldosterone deficiency; 18-hydroxylase deficiency; corticosterone methyl oxidase [CMO] type 1 deficiency; aldosterone synthase deficiency)

| | |
|---|---|
| MIM | 203400 |
| Clinical features | Infants present with signs and symptoms typical of mineralocorticoid deficiency, including failure to thrive, dehydration, hypernatremia, and hypokalemia. Intermittent fever is also common. Only mutations that severely impair enzymatic function will be expressed clinically. Patients respond well to mineralocorticoid replacement. |
| Primary tissue or gland affected | Adrenal cortex. |
| Age of onset | Patients present during infancy with dehydration, poor feeding, and failure to thrive. |
| Epidemiology | Rare, with only a few cases reported in the world literature. |
| Inheritance | Autosomal recessive. |
| Chromosomal location | 8q21 |
| Genes | *CYP11B2* (aldosterone synthase). |
| Mutational spectrum | Mutations include a 5-nucleotide deletion that results in premature termination, several missense mutations, and a single base change that creates a premature termination codon. |
| Effect of mutation | Aldosterone synthase catalyzes the two last steps in the synthesis of aldosterone, and thus has two separate enzymatic functions: 18-hydroxylase and 18-oxidase activity. The two enzymatic activities may be affected differently depending on the specific mutation. Therefore, the relative accumulation of the precursors |

corticosterone and 18-OH corticosterone may vary depending on the genetic defect.

**Diagnosis**

The initial biochemical diagnosis is made by documenting low or absent aldosterone levels in the presence of elevated renin and the appropriate clinical picture. The measurement of precursors and their metabolites will precisely define the enzymatic defect. The type 1 defect is expected to produce low plasma concentrations of 18-OH corticosterone and aldosterone, with low urinary excretion of their metabolites. Type 2 deficiency is associated with low aldosterone, but elevated plasma 18-OH corticosterone levels, and increased urinary excretion of the metabolite tetrahydro-18-hydroxy 11-dehydrocorticosterone. Since both reactions are catalyzed by products of the same gene, it is not surprising that, in individual patients, the differentiation of type 1 and type 2 defects may be confusing, and some cases may not fit a rigid classification.

**Counseling issues**

Once the disease mutation is identified, genetic counseling and prenatal diagnosis are possible. Early diagnosis and treatment prevents the metabolic and developmental defects.

**Notes**

Isolated aldosterone deficiency must be differentiated from other, more common adrenal enzyme defects.

**References**

Peter M, Fawaz L, Drop SL et al. Hereditary defect in biosynthesis of aldosterone: Aldosterone synthase deficiency 1964–1997. *J Clin Endocrinol Metab* 1997;82:3525–8.

Portrat-Doyen S, Tourniaire J, Richard O et al. Isolated aldosterone synthase deficiency caused by simultaneous E198D and V386A mutations in the CYP11B2 gene. *J Clin Endocrinol Metab* 1998;83:4156–61.

# Aldosteronism, Sensitive to Dexamethasone

(also known as: glucocorticoid-suppressible hyperaldosteronism [GSH]; glucocorticoid-remediable hyperaldosteronism [GRH]; ACTH-dependent hyperaldosteronism; glucocorticoid-remediable aldosteronism [GRA])

| | |
|---|---|
| MIM | 103900 |
| Clinical features | First described in 1966, this disease presents as primary hyperaldosteronism with hypertension, hypokalemia, elevated aldosterone levels, and suppressed renin. Like patients with aldosterone-producing adenomas, patients with GSH have a paradoxical decline in aldosterone levels upon standing. |
| Primary tissue or gland affected | Adrenal cortex. |
| Age of onset | Clinical diagnosis is in childhood or adulthood. |
| Epidemiology | Only a small number of affected pedigrees have been reported, and the true prevalence of this syndrome is unknown. |
| Inheritance | Autosomal dominant. |
| Chromosomal location | 8q21 |
| Genes | CYP11B2 (cytochrome P450, subfamily XIB, polypeptide 2), CYP11B1 (cytochrome P450, subfamily XIB, polypeptide 1). |
| Mutational spectrum | The syndrome is caused by nonhomologous pairing and unequal crossing over between two contiguous genes. The result is a fusion gene that contains the 5′ regulatory elements of CYP11B1 and the 3′ elements of CYP11B2. The functional result is the expression of aldosterone synthase (CYP11B2) under the control of an ACTH-sensitive promoter instead of the normal angiotensin 2-sensitive promoter. |

**Effect of mutation**
A report of 21 affected descendents of an English convict revealed extraordinary phenotypic heterogeneity. Affected members were often normokalemic and normotensive until late in life.

**Diagnosis**
Primary hyperaldosteronism is diagnosed by demonstrating elevated aldosterone levels that are not suppressed with salt loading, in the presence of suppressed renin levels that are not stimulated by diuresis or a low-salt diet. This syndrome should be suspected in young patients with an appropriate family history and a paradoxical decline in aldosterone levels upon standing. The clinical diagnosis can be made by treatment with dexamethasone for 4–6 weeks, which will result in the suppression of aldosterone secretion.

**Counseling issues**
This disease is transmitted as an autosomal dominant trait, but is readily treatable once the diagnosis is made.

**Notes**
Given the markedly variable phenotype, this syndrome may be more common than previously thought. Increased aldosterone secretion may also be caused by a single-base polymorphism that was shown to alter aldosterone sensitivity to angiotensin 2. The prevalence of this polymorphism, and its contribution to salt-sensitive essential hypertension, is not known.

**References**

MacConnachie AA, Kelly KF, McNamara A et al. Rapid diagnosis and identification of cross-over sites in patients with glucocorticoid remediable aldosteronism. *J Clin Endocrinol Metab* 1998;83:4328–31.

Pascoe L, Jeunemaitre X, Lebrethon MC et al. Glucocorticoid-suppressible hyperaldosteronism and adrenal tumors occurring in a single French pedigree. *J Clin Invest* 1995;96:2236–46.

# Nonclassical Congenital Adrenal Hyperplasia

(also known as: adult-onset congenital adrenal hyperplasia [CAH]; cryptic CAH)

| | |
|---|---|
| **MIM** | 201910 |

**Clinical features**

Patients with nonclassical CAH have mild postnatal hyperandrogenism. The disease is usually caused by a mutation in the gene encoding the 21-hydroxylase enzyme, though mutations in other enzymes of the steroid biosynthetic pathway have also been reported. Males are essentially asymptomatic, whereas females show signs of mild virilization, typically becoming clinically evident after puberty. The external genitalia are entirely normal. Symptoms are extremely variable, and many patients may be entirely asymptomatic. Although some girls may present with premature pubarche, the more typical presentation includes acne, hirsutism, and menstrual irregularities of variable severity, starting during the third or fourth decade of life. Thus, in women, the typical presentation is easily confused with that of idiopathic polycystic ovary syndrome. In boys, the clinical differentiation between mild, simple virilizing CAH and nonclassical CAH may be difficult, since 17-hydroxyprogesterone (17-OHP) levels overlap. Aldosterone synthesis and sodium balance are not compromised. Similarly, cortisol secretion is normal and adequate, even in stress situations. Data regarding adult height are controversial, with some studies showing a lower mean height when compared with the general population.

**Primary tissue or gland affected**

Adrenal cortex.

**Other organs, tissues, or glands affected**

Ovaries.

**Age of onset**

Diagnosis is usually made after puberty, with the onset of hirsutism and menstrual abnormalities.

| | |
|---|---|
| **Epidemiology** | The disease prevalence is estimated at 0.1% in the general population, 1%–2% in Hispanics and Yugoslavs, and 3%–4% in Ashkenazi Jews. |
| **Inheritance** | Autosomal recessive. |
| **Chromosomal location** | 6p21.3 |
| **Genes** | *CYP21A2* (cytochrome P450, subfamily XX1A, polypeptide 2). |
| **Mutational spectrum** | This mild variant of CAH is caused by any of several missense mutations, the most common of which are V281L and P30L. |
| **Effect of mutation** | There is a 20%–50% decrease in 21-hydroxylase activity. |
| **Diagnosis** | The diagnosis is suspected when a young woman has clinical signs of hyperandrogenism. Biochemical diagnosis is based on the documentation of elevated plasma levels of 17-OHP, a precursor of cortisol biosynthesis, before or after ACTH stimulation. Nonclassical CAH patients with 21-hydroxylase mutations typically have ACTH-stimulated 17-OHP levels of between 1000 ng/dL and 10 000 ng/dL. This compares with levels >10 000 ng/dL in classical CAH and <1000 ng/dL in carriers of severe *CYP21A2* mutations. However, there is significant overlap between stimulated 17-OHP levels in asymptomatic carriers of severe mutations and patients homozygous for nonclassical mutations. Treatment is based on suppression of the adrenal axis, usually with dexamethasone 0.25–0.5 mg before bed, or suppression of androgen activity using an androgen antagonist such as spironolactone or cyproterone acetate. |
| **Counseling issues** | The place of molecular diagnosis in nonclassical CAH is unclear. Being a mild disease, and often entirely asymptomatic, it may be more appropriate to consider the mild mutations as polymorphisms, thus avoiding the psychological and social implications of being |

diagnosed with a genetic disease. On the other hand, some patients with nonclassical disease are compound heterozygotes for severe and mild mutations. These patients are at risk of having a child with severe classical CAH if their partner is an asymptomatic carrier of a severe mutation. The decision of whether to pursue molecular diagnosis must take these possibilities into consideration.

**Notes**

Nonclassical CAH may also be caused by mild mutations in other enzymes responsible for steroidogenesis. Mild 11β-hydroxylase deficiency and mild 3β-hydroxysteroid dehydrogenase deficiency have been reported, but are not yet fully characterized.

**References**

White PC, Speiser PW. Congenital adrenal hyperplasia due to 21-hydroxylase deficiency. *Endocr Rev* 2000;21:245–91.

# Adrenal Unresponsiveness to ACTH

(also known as: melanocortin 2 receptor mutation; ACTH receptor mutation;
familial glucocorticoid deficiency [FGD])

| | |
|---|---|
| **MIM** | 202200 |
| **Clinical features** | Primary presenting symptoms are hyperpigmentation, feeding problems, and hypoglycemia. ACTH receptors on mononuclear leukocytes appear to be similar to those in the adrenal cortex. ACTH binding assays using peripheral monocytes may aid in establishing the diagnosis. |
| **Primary tissue or gland affected** | Adrenal cortex. |
| **Age of onset** | Symptoms are usually present at birth. |
| **Epidemiology** | Rare. |
| **Inheritance** | Autosomal recessive. |
| **Chromosomal location** | 18p11.2 |
| **Genes** | *MC2R* (melanocortin 2 receptor). |
| **Mutational spectrum** | Missense, nonsense, and insertion (frame-shift) mutations have been identified. |
| **Effect of mutation** | Mutations result in failure to produce a biologically active protein. |
| **Diagnosis** | The clinical diagnosis is suspected when extremely elevated ACTH levels are found in association with low cortisol, and low levels of cortisol biosythetic precursors. Other forms of ACTH unresponsiveness also exist. Adrenal aplasia, Allgrove syndrome |

(MIM 231550), bilateral adrenal hemorrhage, and biosynthetic defects within the adrenal cortex (see the congenital adrenal hyperplasias) will also cause elevated ACTH with low cortisol, but, in the latter, levels of other adrenal steroids will also be elevated. In some patients with clinical evidence of adrenal unresponsiveness to ACTH, no mutations in *MC2R* were found, and in a few of these, linkage excluded this chromosomal region. Thus, mutations in other genes may cause an identical clinical syndrome, tentatively called familial glucocorticoid deficiency type 2.

**Notes**

MC2R is an alternative name for the ACTH receptor. An apparent excess of males with this syndrome suggests that an X-linked variant may also exist. Apparent ACTH unresponsiveness has been described in families in which linkage to the *MC2R* locus can be excluded. It is therefore suspected that mutations in other genes may cause a similar syndrome. This syndrome should be differentiated from achalasia–Addisonianism–alacrima syndrome (MIM 231550), which may present as apparently isolated adrenal insufficiency.

**References**

Clark AJ, Cammas FM. The ACTH receptor. *Baillieres Clin Endocrinol Metab* 1996;10:29–47 (Review).

Naville D, Weber A, Genin E et al. Exclusion of the adrenocorticotropin (ACTH) receptor (MC2R) locus in some families with ACTH resistance but no mutations of the MC2R coding sequence (familial glucocorticoid deficiency type 2). *J Clin Endocrinol Metab* 1998;83:3592–6.

*Adrenal Unresponsiveness to ACTH*

# Adrenal Hypoplasia, Congenital

(also known as: AHC; X-linked Addison's disease; adrenal hypoplasia, congenital, with hypogonadotropic hypogonadism [AHCH]; cytomegalic adrenocortical hypoplasia)

| | |
|---|---|
| MIM | 300200 |
| Clinical features | Features of adrenal insufficiency include vomiting, feeding difficulties, hyperpigmentation, apnea, hypoglycemia, and cardiovascular collapse. Laboratory analysis shows hypoglycemia, hyponatremia, and hyperkalemia. Hypogonadism is also common, as is progressive high-frequency hearing loss. The latter becomes evident during the second decade of life. |
| Primary tissue or gland affected | Adrenal cortex. |
| Other organs, tissues, or glands affected | Pituitary, gonadotrophs. |
| Age of onset | Symptoms are present at birth. However, if disease is not suspected, AHC may go undiagnosed until later in the first decade of life. |
| Epidemiology | Rare. |
| Inheritance | X-linked. |
| Chromosomal location | Xp21.3-p21.2 |
| Genes | NROB1 (also known as DAX1; encodes a member of the nuclear hormone receptor superfamily). |
| Mutational spectrum | At least 42 different deletion, missense, and nonsense mutations have been described. |

**Effect of mutation**   No functional protein is produced. NROB1 is a member of the nuclear hormone receptor superfamily. It functions as an antitestes gene. Gene dosage is critical, and duplications of the gene in an XY individual result in the development of female genitalia.

**Diagnosis**   On histology, there is a lack of organization of the adrenal cortex, resulting in cords and clumps of large, pale-staining cells. Because of this, the entity is sometimes referred to as cytomegalic adrenocortical hypoplasia.

**Counseling issues**   AHC is transmitted from an asymptomatic female carrier. Male siblings have a 50% chance of inheriting the disease. This is a severe disease that may result in death if not treated. However, life-threatening adrenal insufficiency can be easily treated with glucocorticoid and mineralocorticoid replacement if diagnosed in time.

**Notes**   Patients have been reported with a combination of AHC, glycerol kinase deficiency (MIM 307030), and Duchenne's muscular dystrophy (MIM 310200). This is caused by a deletion disrupting all three genes (*NROB1*, *GK*, and *DMD*), which are located adjacent to one another on chromosome Xp21.

# Adrenoleukodystrophy

(also known as: ALD; Addison's disease and cerebral sclerosis; adrenomyeloneuropathy
[AMN]; Siemerling–Creutzfeldt disease; melanodermic leukodystrophy)

| | |
|---|---|
| MIM | 300100 |

| | |
|---|---|
| Clinical features | The clinical phenotype is very variable. Approximately 75% of ALD patients have adrenocortical insufficiency, which is most often diagnosed concurrently with the neurologic disease, but may appear before or after the onset of neurologic symptoms. About 80% of patients present during childhood with the severe, cerebral form of the disease, which is associated with a marked inflammatory component. Presenting symptoms include behavioral changes, attention deficit disorder, and impaired vision and hearing. Spasticity, seizures, and dementia appear soon after, and these progress to a vegetative state and death, usually within 2–3 years of the initial diagnosis. Hypogonadism is common. Hypotension, electrolyte disorders, and hyperpigmentation suggest the presence of Addison's disease. All patients should be screened and followed for the development of adrenal insufficiency, since this can be fatal if not diagnosed and treated. Some patients may present later in life with adrenomyeloneuropathy without apparent cerebral symptoms. Although this form of the disease was initially thought to be 'benign', progression to the severe, cerebral form has been reported in 20% of cases. Occasionally, initial presentation may be limited to adrenal insufficiency without any noticible neurologic deficit (Addison-only phenotype); however, as many as 50% of these patients will eventually develop neurologic symptoms. Some patients with gene mutations are entirely asymptomatic. The presence of modifier genes has been proposed as an explanation for this extreme clinical variability. Heterozygous females may have mild to moderate spastic paraparesis or aldosterone deficiency; however, glucocorticoid insufficiency rarely occurs. Aldosterone deficiency appears to be exacerbated by the use of nonsteroidal anti-inflammatory drugs. |

| | |
|---|---|
| **Primary tissue or gland affected** | Adrenal cortex. |
| **Other organs, tissues, or glands affected** | Gonads. |
| **Age of onset** | The age of onset is variable, with about 80% of patients presenting in childhood (mean: approximately 7 years old). Adolescent and adult-onset forms also exist. |
| **Epidemiology** | The estimated birth incidence is approximately 1:42 000 males. This may be an underestimation due to underdiagnosis of mild or atypical disease. The incidence in females is essentially zero, since most affected males are infertile. |
| **Inheritance** | X-linked. |
| **Chromosomal location** | Xq28 |
| **Genes** | *ABCD1* (ATP-binding cassette, subfamily D, member 1). |
| **Mutational spectrum** | Missense, nonsense, insertion, and deletion mutations have been reported. More than 400 different *ABCD1* mutations have been identified, and these are recorded on an international registry database accessible through the website: http://www.x-ald.nl. |
| **Effect of mutation** | ABCD1 codes for the adrenoleukodystrophy protein, designated ALDP, which is involved in the transport or anchoring of very long-chain fatty acid (VLCFA)-coenzyme A synthetase. Mutations prevent the formation of functionally active protein and result in the accumulation of VLCFA in serum and tissues. |
| **Diagnosis** | The diagnosis is suspected when male patients present with typical neurologic symptoms and/or new-onset adrenocortical insufficiency. The biochemical diagnosis is based on finding elevated plasma |

levels of VLCFA. Mutation analysis will confirm the diagnosis; however, given the size of the gene, this is a difficult undertaking. Prenatal diagnosis can be made genetically if the specific mutation in the family is known, or can be made by measuring VLCFA levels in cultured amniocytes or chorionic villus cells. Heterozygous females have VLCFA levels intermediate between normal levels and those seen in affected males. Cultured fibroblasts show two distinct types of clones: those with elevated VLCFA levels, and normal cells. Subclinical cortisol or aldosterone deficiency can be identified by testing the response of both hormones to ACTH stimulation.

**Counseling issues**

Bone marrow transplantation may be beneficial when performed at an early stage of the disease. Other treatments under investigation include lovastatin, fenofibrate, and butyric acid analogs, such as 4-phenylbutyrate. Gene therapy to correct the VLCFA defect has been attempted. After an index case is identified in a family, prenatal diagnosis is possible even if the specific mutation is not known.

**Notes**

This disease is portayed in the movie *Lorenzo's Oil,* which is based on an apparently true story and describes the family of an affected child who were convinced that they had discovered an effective dietary treatment approach. However, controlled studies using the treatment proposed in this movie have failed to demonstrate any significant clinical improvement.

**References**

Raymond GV. Peroxisomal disorders. *Curr Opin Neurol* 2001;14:783–7.

Van Geel BM, Bezman L, Loes DJ et al. Evolution of phenotypes in adult male patients with X-linked adrenoleukodystrophy. *Ann Neurol* 2001;49:186–94.

# Achalasia–Addisonianism–Alacrima Syndrome

(also known as: triple-A [AAA] syndrome; Allgrove syndrome; glucocorticoid deficiency and achalasia; hypoadrenalism with achalasia)

| | |
|---|---|
| **MIM** | 231550 |
| **Clinical features** | First described in 1978, AAA consists of adrenal insufficiency due to ACTH resistance, achalasia, and alacrima. Autonomic and/or peripheral neuropathy, hyperkeratosis, delayed wound healing, mental retardation, and short stature may also be present. Addison's disease is a necessary component of this syndrome, which may include any combination of the other major clinical findings. Mineralocorticoid deficiency has been reported, but is usually not present. The onset of clinical symptoms may not be simultaneous, and some patients with apparently isolated ACTH resistance have been subsequently diagnosed with AAA. Postmortem studies reveal the absence of the zona fasciculata, with an almost normal zona glomerulosa. The mutations are clearly recessive, since none of the heterozygous parents or siblings fulfil any of the criteria for the syndrome. |
| **Primary tissue or gland affected** | Adrenal cortex. |
| **Age of onset** | Diagnosis is typically made during childhood, usually before 10 years of age, and frequently in association with acute adrenal insufficiency, which may be precipitated by an intercurrent illness. Other clinical findings, such as alacrima, may precede adrenal insufficiency. One case of adult diagnosis (age 35 years) has been reported. |
| **Epidemiology** | The disease is rare, with apparent founder mutations, and hence an increased incidence in Puerto Rican and North African populations. |
| **Inheritance** | Autosomal recessive. |
| **Chromosomal location** | 12q13 |

| | |
|---|---|
| **Genes** | *AAAS* (aladin). |
| **Mutational spectrum** | Truncation and splice-site mutations have been described. A single splice-donor site founder mutation was found in North African families, and appears to date back >2400 years. |
| **Effect of mutation** | Mutations prevent or modify the expression of the protein aladin, which belongs to the WD-repeat family of regulatory proteins. The function of the protein is not yet known. Based on its structure, it is postulated to be involved either in cytoplasmic trafficking or peroxisomal activities. Clinical severity varies in patients with identical mutations. |
| **Diagnosis** | The clinical diagnosis is based on the finding of ACTH-resistant adrenal insufficiency combined with achalasia and/or alacrima. Because different aspects of the syndrome may appear at different ages, family studies and follow-up are required to exclude the diagnosis in a patient with apparently isolated ACTH resistance. Genetic diagnosis is definitive, though not all mutant alleles can be identified by studying the coding sequence. In a recent study, a patient was identified who was heterozygous for a single mutation. The coding sequence and splice sites were normal on the second allele, as was long-range polymerase chain reaction. However, the heterozygous parent and siblings were asymptomatic, suggesting that the second allele carried an unidentified mutation outside the coding region. The combination of adrenal insufficiency and neurologic symptoms may suggest the diagnosis of X-linked adrenoleukodystrophy (300100), which is clearly a different entity. |
| **Counseling issues** | Definitive clinical and genetic diagnosis of affected probands will provide the opportunity for genetic counseling and prenatal diagnosis. Prompt and efficient treatment of adrenal insufficiency may prevent or minimize the mental retardation and short stature that are sometimes seen. |

**Notes**

Although some patients initially present with isolated ACTH-resistant adrenal insufficiency (iACTHR), most will show additional signs and symptoms of the syndrome. A cohort of patients with iACTHR with no demonstrable mutation on the ACTH receptor was screened for mutations in the *AAAS* gene; none were found, suggesting that iACTHR is not a *forme fruste* of AAA.

**References**

Bentes C, Santos-Bento M, de Sa J et al. Allgrove syndrome in adulthood. *Muscle Nerve* 2001;24:292–6.

Sandrini F, Farmakidis C, Kirschner LS et al. Spectrum of mutations of the AAAS gene in Allgrove syndrome: lack of mutations in six kindreds with isolated resistance to corticotropin. *J Clin Endocrinol Metab* 2001;86:5433–7.

# Nephrogenic Diabetes Insipidus

(also known as: nDI)

| | |
|---|---|
| **MIM** | 304800 (X-linked), 222000 (autosomal recessive), 125800 (autosomal dominant) |
| **Clinical features** | The clinical symptoms of congenital nDI are similar regardless of the underlying mutation, with the exception of three mutations of the V2 receptor (D85N, G201D, and P322S). These mutations are associated with a milder form of nDI, with presenting symptoms appearing at age 10 years, and no growth retardation. For all other mutations, affected infants have a poor appetite and suffer from recurrent fever, vomiting, dehydration, and hypernatremia. Failure to thrive, developmental delay, and mental retardation may occur if the diagnosis is delayed and adequate water is not provided. In severe cases in adult patients, urine production may exceed 20 L per 24 hours. Late sequelae may include hydronephrosis, hydrostatic nephropathy, and renal insufficiency. |
| **Primary tissue or gland affected** | Kidney. |
| **Age of onset** | Acquired nDI can appear at any age. In congenital nDI, the renal concentration defect is present soon after birth. Symptoms appearing in the first year of life are often nonspecific (poor appetite, vomiting, growth retardation), and diagnosis may be delayed until later in childhood when polyuria and polydipsia are recognized. |
| **Epidemiology** | Congenital nDI has a prevalence of 1:250 000 males. X-linked nDI is found in 90%, while autosomal recessive inheritance is noted in 10%. Autosomal dominant nDI is very rare. Acquired forms of mild nDI are common, especially among elderly patients, with variable degrees of renal failure. |
| **Inheritance** | X-linked, autosomal recessive, autosomal dominant. |

| Chromosomal location | Xq28 (AVPR2), 12q13 (AQP2). |
|---|---|
| Genes | *AVPR2* (vasopressin receptor 2), *AQP2* (aquaporin 2). |
| Mutational spectrum | More than 180 mutations of the *AVPR2* gene have been identified as causing nDI, including nucleotide deletions, insertions, and substitutions. Mutations of *AQP2* in families with autosomal recessive nDI include missense, nonsense, and nucleotide deletions. A point mutation resulting in the substitution of lysine for glutamic acid at position 258 of *AQP2* was identified in a family with autosomal dominant nDI. |
| Effect of mutation | *AVPR2* mutations have been classified into four subgroups based on *in vitro* studies. Class I mutations result in abnormal splicing, frame shifts, and premature termination of translation. The truncated protein resulting from these mutations leads to a complete lack of cell surface receptors. Most *AVPR2* mutations (missense, in-frame deletions, and nonsense) belong to class II: protein translation is complete, but because of abnormal folding or assembling, the protein is trapped in the endoplasmic reticulum and undergoes degradation in the proteasomes. Receptor expression on the cell surface is either absent or severely reduced. In class III mutations, cell surface receptors are present in normal abundance, but coupling to the intracellular-stimulating G protein is impaired. Class IV mutations are characterized by reduced receptor affinity for arginine vasopressin (AVP) binding. Most of the mutations that cause autosomal recessive forms of nDI result in abnormal folding of the AQP2 protein, leading to impaired export from the endoplasmic reticulum. Other *AQP2* mutations cause nonfunctional water channels. In autosomal dominant nDI, the mutant protein forms heterotetramers with wild-type AQP2, and prevents transport of wild-type AQP2 to the plasma membrane. |

| | |
|---|---|
| **Diagnosis** | Determination of urine volume, measurement of urine and serum osmolality, and water deprivation tests to diagnose DI are described in the section on central DI (see p. 52). Failure to increase urine concentration after 1-desamino-8-D-(DD)AVP administration confirms the diagnosis of nDI. Extrarenal responses to DDAVP (increases in von Willebrand factor, factor VIII, and tissue-type plasminogen activator levels) are normal in autosomal nDI, but absent in patients with X-linked nDI. Causes of acquired nDI include hypercalcemia, potassium depletion, chronic renal disease, sickle-cell anemia, and drug effects (most often due to lithium or demeclocycline). |
| **Counseling issues** | Early diagnosis and treatment are essential in order to prevent failure to thrive and mental retardation. |
| **Notes** | The recommended treatment for nDI includes adequate fluid intake, along with a combination of hydrochlorothiazide and either amiloride or indomethacin. Recent *in vitro* studies have identified pharmacological 'chaperones' that rescue AVPR2 or AQP2 mutants trapped in the endoplasmic reticulum and facilitate their transport to the plasma membrane. |
| **References** | Knoers NV, Deen PM. Molecular and cellular defects in nephrogenic diabetes insipidus. *Pediatr Nephrol* 2001;16:1146–52 (Review).

Oksche A, Rosenthal W. The molecular basis of nephrogenic diabetes insipidus. *J Mol Med* 1998;76:326–37 (Review). |

# Syndrome of Apparent Mineralocorticoid Excess

(also known as: AME; cortisol 11β-ketoreductase deficiency; hydroxysteroid dehydrogenase [HSD]11, type 2)

| | |
|---|---|
| **MIM** | 218030 |
| **Clinical features** | Clinical manifestations include salt-sensitive hypertension, hypokalemia, low plasma renin activity, and low serum aldosterone levels. Hypertension responds to mineralocorticoid receptor (MR) blockade with spironolactone. The clinical and electrolyte findings suggest mineralocorticoid excess; however, aldosterone and other mineralocorticoid levels are low. The urinary cortisol to cortisone ratio is increased secondary to the enzyme block. The clinical findings are caused by abnormal activation of the MR by intracellular cortisol. |
| **Primary tissue or gland affected** | Kidney. |
| **Other organs, tissues, or glands affected** | Adrenal cortex. |
| **Age of onset** | Clinical diagnosis is usually made during childhood, sometimes presenting with chronic severe hypertension, including retinopathy and stroke. Milder cases have been reported to present in adulthood, and are presumably due to mutations that result in a less severe decrease in enzyme function. |
| **Epidemiology** | This is a rare syndrome: 25 kindreds with proven gene mutations have been identified. The overall incidence is estimated to be <1:250 000 in white people. |
| **Inheritance** | Autosomal recessive. |
| **Chromosomal location** | 16q22 |

| | |
|---|---|
| **Genes** | *11BHSD2* (11β-dehydrogenase type II). |
| **Mutational spectrum** | At least 20 different mutations have been reported. Most are point mutations, resulting in amino acid changes. |
| **Effect of mutation** | The mutation abolishes or markedly reduces enzyme activity. *In vitro*, the MR is sensitive to both cortisol and aldosterone, but not to cortisone. Cells that express the MR also express high levels of 11β-HSD2, which rapidly converts cortisol to cortisone. Thus, under normal circumstances, circulating cortisol levels are unable to activate the receptor, since intracellular cortisol levels are negligible. However, in the absence of enzyme activity, intracellular cortisol levels are elevated, and activate the receptor even in the absence of aldosterone. |
| **Diagnosis** | The diagnosis is suspected when clinical and electrolyte studies suggest mineralocorticoid excess, but blood levels of aldosterone and other mineralocorticoids are decreased. Hypertension responds to treatment with the MR antagonist spironolactone, excluding a receptor mutation. Definitive diagnosis is made by identifying a mutation in the responsible gene. |
| **Counseling issues** | This is a rare, but treatable, cause of juvenile-onset hypertension. Early and aggressive treatment will prevent the long-term sequelae of hypertension and electrolyte imbalance. Thus, genetic counseling and testing are warranted whenever a family with the mutation is identified. |
| **Notes** | Glycyrrhetinic acid, a component of licorice, reversibly and specifically inhibits 11β-HSD2, and will therefore cause a clinical syndrome identical to that produced by enzyme mutations. Activity of 11β-HSD2 may play a role in the pathogenesis of salt-sensitive essential hypertension and in pre-eclampsia. |
| **References** | Ferrari P, Bianchetti M, Frey FJ. Juvenile hypertension, the role of genetically altered steroid metabolism. *Horm Res* 2001;55:213–23 (Review). |

# Liddle Syndrome

(also known as: pseudoaldosteronism; pseudohyperaldosteronism)

| | |
|---|---|
| **MIM** | 177200 |

**Clinical features**  This syndrome was first described in 1963, and is characterized by severe hypertension, hypokalemia, and metabolic alkalosis that mimics primary hyperaldosteronism. Mutations in the β or γ subunits of the renal epithelial sodium channel (ENaC) cause an increased channel current, resulting in greater sodium reabsorption. Increased intracellular sodium creates a transepithelial potential difference that drives potassium secretion through apically located channels. Volume expansion caused by constitutive sodium reabsorption results in hypertension, with suppression of renin and aldosterone secretion. Neither the hypokalemia nor hypertension respond to the aldosterone inhibitor spironolactone, but treatment with the potassium-sparing diuretic triamterene does normalize blood pressure and correct the hypokalemia.

**Primary tissue or gland affected**  Kidney.

**Age of onset**  Hypertension has been noted at as early as 10 months of age. The clinical diagnosis is typically delayed if the disease is not suspected due to family history.

**Epidemiology**  Liddle syndrome is very rare, and its incidence and prevalence have not been accurately determined.

**Inheritance**  Autosomal dominant.

**Chromosomal location**  16p13-p12

**Genes**  *SCNN1B* (epithelial sodium channel β), *SCNN1G* (epithelial sodium channel γ).

**Mutational spectrum**      Truncation and missense mutations have been identified on the β subunit of ENaC. The missense mutations are all located in a specific proline-rich region of the molecule that is thought to suppress channel activity under normal circumstances. A truncation mutation in the γ subunit was identified in a single kindred.

**Effect of mutation**      Mutations result in increased activity of the sodium channel. C-terminal truncation of either the β or γ subunit will have the same functional result, suggesting that these subunits are primarily involved in the negative regulation of channel activity.

**Diagnosis**      The diagnosis is suspected when abnormalities suggestive of hyperaldosteronism are found in the presence of suppressed renin and aldosterone levels. Other causes of 'pseudohyperaldosteronism', such as ingestion of large amounts of licorice and the syndrome of apparent mineralocorticoid excess (MIM 218030, previous entry), must be excluded. The diagnosis can be confirmed by identifying an activating mutation in either of the two subunits of ENaC.

**Counseling issues**      Early diagnosis will lead to early therapy and an improved overall prognosis.

**Notes**      Polymorphisms that cause a more subtle modification of channel activity may be important in other forms of low renin hypertension. Several studies have suggested an association between some of these polymorphisms and hypertension in specific populations.

**References**      Grunder S. Liddle's syndrome and pseudohypoaldosteronism type 1. In: Lehmann-Horn F, Jurkat-Rott K, editors. *Channelopathies: Common Mechanisms in Aura, Arrhythmia and Alkalosis.* Amsterdam: Elsevier, 2000:227–98.

Snyder PM. The epithelial Na(+) channel: Cell surface insertion and retrieval in Na(+) homeostasis and hypertension. *Endocr Rev* 2002;23:258–75.

# Bartter's Syndrome

**(also known as: hypokalemic alkalosis with hypercalciuria)**

| | |
|---|---|
| **MIM** | 241200, 601678 (antenatal variant) |

**Clinical features**
Bartter's syndrome is a set of closely related disorders. The primary clinical characteristics include hypokalemia, hypochloremic alkalosis, increased urinary excretion of potassium and prostaglandins (prostaglandin $E_2$), and hyperreninemic hyperaldosteronism (secondary hyperaldosteronism) with normal blood pressure. The classical syndrome usually presents in childhood with failure to thrive. A variant of the disease, known as the antenatal hypercalciuric variant, is characterized by hydramnios, prematurity, and dehydration at birth. Congenital sensorineural deafness may also occur. Hypercalciuria is commonly found, apparently caused by the inhibition of paracellular reabsorption of calcium, and may result in nephrocalcinosis and renal damage. Typically, magnesium loss is minimal and hypomagnesemia is not found.

**Primary tissue or gland affected**
Kidney.

**Other organs, tissues, or glands affected**
Adrenal cortex.

**Age of onset**
The disease is usually diagnosed in infancy, childhood, or early adolescence.

**Epidemiology**
Bartter's syndrome has been reported in all ethnic groups. The incidence has been estimated at about 1.2:1 000 000.

**Inheritance**
Autosomal recessive.

**Chromosomal location**
1p36, 15q15-q21.1, 11q24, 1p31.

| | |
|---|---|
| **Genes** | *CLCNKB* (kidney chloride channel), *SLC12A1* (also known as *NKCC2*, sodium–potassium–chloride transporter 2), *ROMK* (potassium ion channel), *KCNJ1* (potassium channel), *BSND* (Barttin). |
| **Mutational spectrum** | Loss of function mutations in each of the above-mentioned genes have been reported in some, but not all, patients with Bartter's syndrome. Deletion, missense, and truncation mutations have been described. |
| **Effect of mutation** | Mutations result in loss of function of the affected gene product. Defects in any of these genes will impair the net reabsorption of sodium chloride in the thick ascending limb of Henle's loop, thereby causing the delivery of excessive amounts of sodium to the distal nephron, and resulting in salt wasting, volume contraction, and secondary hyperaldosteronism. This results in hypokalemic alkalosis and stimulates the production of prostaglandin $E_2$, another hallmark of the syndrome. |
| **Diagnosis** | The clinical diagnosis is suggested by finding the appropriate electrolyte disorder (hypokalemic metabolic alkalosis) associated with normal blood pressure. Differential diagnoses for hypokalemic metabolic alkalosis include: primary mineralocorticoid excess, which is excluded by the absence of hypertension; and secondary hyperaldosteronism due to extrarenal fluid loss, which is excluded by the finding of high urinary chloride excretion. Mutation analysis can provide a definitive diagnosis; however, some cases of Bartter's syndrome are apparently caused by mutations in as yet unidentified genes. |
| **Counseling issues** | Early diagnosis and treatment can significantly improve the prognosis. |
| **Notes** | Similar, but not identical, electrolyte disturbances are seen in Gitelman's syndrome (MIM 263800, next entry), which is usually much milder, diagnosed later in life, and has a different genetic etiology. |

**References**

Birkenhager R, Otto E, Schurmann MJ et al. Mutation of BSND causes Bartter syndrome with sensorineural deafness and kidney failure. *Nat Genet* 2001;29:310–4.

Scheinman SJ, Guay-Woodford LM, Thakker RV et al. Genetic disorders of renal electrolyte transport. *N Engl J Med* 1999;340:1177–87 (Review).

# Gitelman's Syndrome

(also known as: Gitelman's variant of Bartter's syndrome; primary renotubular hypomagnesemia–hypokalemia with hypocalciuria)

| | |
|---|---|
| **MIM** | 263800 |
| **Clinical features** | The typical findings are hypokalemia, hypomagnesemia, alkalosis, hypovolemia, and hypocalciuria. Increased net renal sodium loss results in secondary hyperaldosteronism (elevated renin and aldosterone), thus exacerbating the hypokalemia. One third of patients have short stature. |
| **Primary tissue or gland affected** | Kidney. |
| **Other organs, tissues, or glands affected** | Adrenal cortex. |
| **Age of onset** | Diagnosis is usually delayed until late childhood or adulthood. |
| **Epidemiology** | The prevalence of heterozygotes is estimated at about 1%. |
| **Inheritance** | Autosomal recessive. |
| **Chromosomal location** | 16q13 |
| **Genes** | *SLC12A3* (also known as *TSC* [thiazide-sensitive NaCl cotransporter]). |
| **Mutational spectrum** | Mutations include missense, termination, and splice-site mutations, as well as deletions. |
| **Effect of mutation** | Mutations result in a lack of functional protein. As with other specific renal transport defects, primary defects in the *TSC* gene result in |

secondary changes in other transport mechanisms, and thus several additional electrolyte disturbances. Mutations result in defective reabsorption of NaCl in the early distal tubule of the kidney. This results in increased calcium and sodium reabsorption, and increased potassium excretion in the more distal portion of the tubule. Net increased NaCl excretion results in hypovolemia, stimulating renin and causing secondary hyperaldosteronism, and thus further increasing potassium excretion as well as stimulating $H^+$ and magnesium secretion.

**Diagnosis**

The diagnosis is suspected by demonstrating the appropriate electrolyte abnormalities, and is confirmed by direct mutation analysis. Patients are frequently asymptomatic, or may present with transient episodes of weakness, abdominal pain, fever, or tetany. Because of the nonspecific, transient symptoms, diagnosis may be delayed until adulthood. Gitelman's syndrome can be differentiated from Bartter's syndrome (MIM 241200) since it is clinically milder and the electrolyte disturbance is somewhat different, particularly the hypocalciuria and hypomagnesemia, which is typical of this syndrome and not found in Bartter's sydrome.

**Counseling issues**

Family members of affected individuals should be screened for relevant electrolyte abnormalities, and treatment should be instituted to correct abnormalities, particularly hypokalemia.

**Notes**

See also Bartter's Syndrome (MIM 241200, previous entry), a similar clinical syndrome caused by a mutation in a different gene.

**References**

Barakat AJ, Rennert OM. Gitelman's syndrome (familial hypokalemia–hypomagnesemia). *J Nephrol* 2001;14:43–7.

Schmidt H, Kabesch M, Schwarz HP et al. Clinical, biochemical and molecular genetic data in five children with Gitelman's syndrome. *Horm Metab Res* 2001;33:354–7.

# Pseudohypoaldosteronism Type 1

(also known as: PHA1)

| | |
|---|---|
| **MIM** | 264350 (autosomal recessive), 177735 (autosomal dominant) |
| **Clinical features** | Pseudohypoaldosteronism type 1 is characterized by renal sodium wasting with high sodium concentrations in sweat, saliva, and stool. Hyperkalemia is caused by decreased renal potassium excretion in the presence of elevated renin and aldosterone levels. The disease usually presents in the neonatal period with vomiting, hyponatremia, and failure to thrive. Respiratory infections are common and appear to be caused by abnormal sodium channel activity in the respiratory epithelium. Patients with the autosomal dominant form have a similar disorder of renal electrolyte handling, but do not have increased respiratory infections, since the respiratory sodium channels function normally. Interestingly, in patients with the dominant form of the disease, carbenoxolone, which inhibits 11β-hydroxysteroid dehydrogenase (the enzyme responsible for the local inactivation of cortisol by conversion to cortisone), partially corrects the mineralocorticoid resistance. Locally elevated cortisol levels activate the wild-type receptor, suggesting that the clinical disease is caused by haploinsufficiency of the normal allele. |
| **Primary tissue or gland affected** | Kidney. |
| **Other organs, tissues, or glands affected** | Sweat glands, respiratory epithelium. |
| **Age of onset** | Neonatal. |
| **Epidemiology** | The syndrome is rare, and the precise incidence has not been determined. |
| **Inheritance** | Autosomal recessive, autosomal dominant. |

| | |
|---|---|
| **Chromosomal location** | 16p13-p12, 12p13, 4q31.1 |
| **Genes** | *SCNN1A* (epithelial sodium channel [ENaC]-α), *SCNN1B* (ENaC-β), *SCNN1G* (ENaC-γ), *NR3C2* (mineralocorticoid [aldosterone] receptor). |
| **Mutational spectrum** | Recessive mutations in each of the three subunits of ENaC have been described. Four different dominant mutations have been reported in the mineralocorticoid receptor. |
| **Effect of mutation** | Mutations in any of the three subunits of ENaC result in loss of function of the channel, and thus a clinically identical syndrome. Mineralocorticoid receptor mutations create a nonfunctional receptor. Clinical disease may be caused by haploinsufficiency of the wild-type receptor. |
| **Diagnosis** | The diagnosis is based on finding the appropriate electrolyte abnormalities (hyperkalemia, hyponatremia) in the presence of elevated renin and aldosterone levels. Recurrent respiratory infections, which are typically seen in patients with ENaC defects, may be mistaken for cystic fibrosis. Patients with aldosterone receptor defects present with salt wasting and hyperkalemia, but without pulmonary or other organ system involvement. |
| **Counseling issues** | Prenatal genetic diagnosis is possible for subsequent siblings if the precise mutation is identified in the proband. Early diagnosis and aggressive treatment are associated with improved survival. |
| **Notes** | With aggressive salt replacement and control of hyperkalemia, survival is possible. The electrolyte disorder appears to become less severe with age, particularly in patients with mineralocorticoid receptor defects. See also Liddle's syndrome (MIM 177200, p. 272), which is caused by gain of function mutations in the same sodium channel. |

**References**

Scheinman SJ, Guay-Woodford LM, Thakker RV et al. Genetic disorders of renal electrolyte transport. *N Engl J Med* 1999;340:1177–87.

Snyder PM. The epithelial Na(+) channel: Cell surface insertion and retrieval in Na(+) homeostasis and hypertension. *Endocr Rev* 2002;23:258–75 (Review).

# Pseudohypoaldosteronism Type 2

(also known as: PHA2; familial hyperkalemia and hypertension;
Gordon hyperkalemia–hypertension syndrome)

| | |
|---|---|
| **MIM** | 145260 |

**Clinical features**  PHA2 is characterized by hyperkalemia and hypertension, despite normal glomerular filtration and adrenal function. Three genetically independent subtypes have been described, and are designated PHA2A, B, and C. Hyperchloremic acidosis may be seen, as can secondary hyperkalemic periodic paralysis. A low-salt diet improves the hypertension and decreases urinary sodium, but does not correct hyperkalemia. Similarly, exogenous aldosterone prevents renal sodium loss, but does not improve the hyperkalemia. The electrolyte disturbance is treated with thiazide diuretic. Renin and aldosterone levels are low or inappropriately low for the degree of hyperkalemia. Short stature is frequently seen and appears to be secondary, since early treatment results in catch-up growth. Mental retardation and skeletal abnormalities have also been reported.

**Primary tissue or gland affected**  Adrenal cortex.

**Age of onset**  The age of diagnosis is extremely variable and ranges from 2 weeks to >50 years old. Hyperkalemia appears to precede hypertension.

**Epidemiology**  Over 100 sporadic cases and families have been reported from different ethnic groups. The marked genetic heterogeneity of the cases suggests that each subtype is very rare.

**Inheritance**  Autosomal dominant.

**Chromosomal location**  1q31-q42 (PHA2A), 17q21-q22 (PHA2B, PRKWNK4), 12p13.33 (PHA2C, PRKWNK1).

| Genes | *PRKWNK4* (protein kinase, lysine-deficient 4), *PRKWNK1* (protein kinase, lysine-deficient 1). The responsible gene at locus 1q31-q42 has not yet been identified. |
|---|---|

**Genes**     *PRKWNK4* (protein kinase, lysine-deficient 4), *PRKWNK1* (protein kinase, lysine-deficient 1). The responsible gene at locus 1q31-q42 has not yet been identified.

**Mutational spectrum**     Several missense mutations have been described in the *PRKWNK4* gene, and a 41-kb deletion in the *PRKWNK1* gene.

**Effect of mutation**     The molecular etiologies of PHA2B and PHA2C have been determined, and the disease is caused by loss of function mutations in two different serine-threonine kinases. WNK1 is expressed in most tissues studied, including the distal nephron, whereas WNK4 appears to be expressed primarily in the distal convoluted tubule and the cortical collecting duct.

**Diagnosis**     The diagnosis is suggested in patients with hypertension and hyperkalemia in the absence of any renal damage or decreased glomerular filtration. Renin levels are low, while aldosterone levels are within the normal range, but inappropriately low for the level of hyperkalemia. A marked response to thiazide diuretics is the clinical hallmark of PHA2.

**Counseling issues**     Early diagnosis and thiazide therapy will improve the prognosis, including the short stature frequently seen in untreated patients.

**Notes**     The molecular etiology of PHA2A is still unknown. Analysis of pedigrees has suggested that mutations at additional loci may cause a clinically similar syndrome.

**References**     Achard JM, Disse-Nicodeme S, Fiquet-Kempf B et al. Phenotypic and genetic heterogeneity of familial hyperkalaemic hypertension (Gordon syndrome). *Clin Exp Pharmacol Physiol* 2001;28:1048–52.

Wilson FH, Disse-Nicodeme S, Choate KA et al. Human hypertension caused by mutations in WNK kinases. *Science* 2001;293:1107–12.

# Carney Complex

(also known as: CNC; primary pigmented nodular adrenocortical disease [PPNAD]; myxoma–adrenocortical dysplasia syndrome; nevi, atrial myxoma, mucinosis of skin, endocrine hyperactivity [NAME] syndrome)

| | |
|---|---|
| MIM | 160980 |
| Clinical features | The clinical features of CNC are similar to those seen in McCune–Albright syndrome (MIM 174800, p. 12). Endocrine involvement includes primary adrenocortical nodular dysplasia (ACTH-independent hypercortisolism) and pituitary adenoma (prolactinoma, acromegaly, or nonfunctioning). A classical presentation of hypercortisolism (Cushing's syndrome) is noted in only 25% of patients; most have mild, subclinical, or periodic Cushing's syndrome. Nonendocrine tumors include atrial myxoma, hemangioma, myxoid liposarcoma, neurofibromata, nasopharyngeal schwannoma, virilizing adrenocortical carcinoma, and large-cell calcifying Sertoli-cell tumor of the testes. Cutaneous manifestations include profuse pigmented skin lesions, nevi, ephelides (freckles), lentigines, and labial pigmentation. |
| Primary tissue or gland affected | Adrenal cortex. |
| Age of onset | Heterogeneous. |
| Epidemiology | CNC is exceedingly rare, and has been estimated to be responsible for no more than 1% of clinical Cushing's syndrome cases. Based on this, its incidence can be estimated at about 0.02:100 000/year. Approximately two thirds are familial cases. |
| Inheritance | Autosomal dominant. |
| Chromosomal location | 17q22-q24, 2p16 |

| Genes | PRKAR1A (protein kinase A regulatory subunit α); a further chromosome 2 gene is unknown. |
|---|---|
| Mutational spectrum | Missense mutations and microdeletions have been described. A 2-base pair deletion in exon 4B appears to be due to a mutation hotspot, since it was identified in four unrelated families. Tumors show loss of heterozygosity at this region of chromosome 17, suggesting that the tumors are caused by a double-hit mechanism. In a patient heterozygous for the mutation, somatic loss of the normal allele in a precursor cell results in malignant differentiation. |
| Effect of mutation | Tumors resected from patients with PRKAR1A mutations showed increased cAMP-stimulated protein kinase A (PKA) activity, suggesting that this gene acts as a tumor suppressor gene by modulating the response of PKA to cAMP. |
| Diagnosis | Until recently, diagnosis was based purely on identification of the appropriate symptom complex. The recent discovery of PRKAR1A on chromosome 17 has made molecular diagnosis possible for cases with disease linked to that chromosome. |
| Notes | One family with cardiac myxoma and no other central nervous system stigmata was found to carry a mutation in PRKAR1A, suggesting that familial cardiac myxoma may be a variant of CNC. |
| References | Carney JA, Gordon H, Carpenter PC et al. The complex of myxomas, spotty pigmentation, and endocrine overactivity. *Medicine (Baltimore)* 1985;64:270–83. |
| | Kirschner LS, Carney JA, Pack SD et al. Mutations of the gene encoding the protein kinase A type I-alpha regulatory subunit in patients with the Carney complex. *Nat Genet* 2000;26:89–92. |

# Adrenocortical Carcinoma, Hereditary

(also known as: ADCC)

| | |
|---|---|
| **MIM** | 202300 |
| **Clinical features** | The primary clinical finding is virilization due to increased secretion of steroid precursors with androgenic activity. |
| **Primary tissue or gland affected** | Adrenal cortex. |
| **Age of onset** | Childhood. |
| **Epidemiology** | This is a rare syndrome, representing about 0.4% of childhood tumors. An increased frequency has been reported in patients with Beckwith–Wiedemann syndrome (MIM 130650, p. 219). There is an apparent increased incidence in southern Brazil. |
| **Inheritance** | Autosomal recessive. |
| **Chromosomal location** | 11p15.5 |
| **Genes** | Unknown. |
| **Mutational spectrum** | Specific genomic mutations have not been identified. However, loss of heterozygosity or imprinting abnormalities have been found in the tumors themselves. |
| **Effect of mutation** | Loss of heterozygosity of 11p15.5 in the tumor. |
| **Diagnosis** | The diagnosis is made by identifying a large adrenal mass in a patient with appropriate clinical findings. |

**Notes**          The region that is suspected to contain the disease gene (11p15.5) includes imprinted genes such as *IGF2*, *H19*, and *P57kip2*, all of which are important in cell growth and regulation of the cell cycle.

**References**     Figueiredo BC, Stratakis CA, Sandrini R. Comparative genomic hybridization analysis of adrenocortical tumors of childhood. *J Clin Endocrinol Metab* 1999;84:1116–21.

Gicquel C, Raffin-Sanson ML, Gaston V et al. Structural and functional abnormalities at 11p15 are associated with the malignant phenotype in sporadic adrenocortical tumors: Study on a series of 82 tumors. *J Clin Endocrinol Metab* 1997;82:2559–65.

Mahloudji M, Ronaghy H, Dutz W. Virilizing adrenal carcinoma in two sibs. *J Med Genet* 1971;8:160–3.

# von Hippel–Lindau Syndrome

(also known as: VHL)

| | |
|---|---|
| **MIM** | 193300 |
| **Clinical features** | Clinical findings in VHL include pheochromocytoma, renal cell carcinoma (RCC), retinal angioma, and central nervous system hemangioblastoma. The syndrome is divided into three major subclasses: type 1 is characterized by hemangioblastomas and RCC; type 2a includes hemangioblastomas and pheochromocytoma; type 2b includes hemangioblastoma, RCC, and pheochromocytoma; and type 2c presents with familial pheochromocytoma. The syndrome typically presents clinically with the diagnosis of pheochromocytoma or hemangioblastoma. Germline VHL mutations exist in 35%–50% of patients with familial or bilateral pheochromocytomas, and 3%–11% of patients with sporadic hemangioblastomas under the age of 50 years. |
| **Primary tissue or gland affected** | Adrenal medulla. |
| **Age of onset** | The mean age of diagnosis is 26 years, with about 97% of mutation carriers presenting with symptoms by the age of 60 years. In patients identified as mutation carriers, routine periodic examinations result in significantly earlier diagnosis of disease. Retinal angioma may be observed in the first decade of life, cerebellar hemangioblastoma during the second, and renal cell carcinoma in the third and fourth decades. |
| **Epidemiology** | Incidence has been estimated at between 1:35 000 and 1:45 500 live births. |
| **Inheritance** | Autosomal dominant. |
| **Chromosomal location** | 3p26-p25 |

| | |
|---|---|
| **Genes** | *VHL* (von Hippel–Lindau syndrome tumor suppressor gene). |
| **Mutational spectrum** | Missense, termination, insertion, and deletion mutations have been described. Type 1 VHL is associated with deletion and truncation mutations. Type 2a is most commonly associated with the missense mutations Tyr98His and Tyr112His, type 2b with other missense mutations, and type 2c with the missense mutations Leu188Val, Val84Leu, and Ser80Leu. As many as 40% of VHL patients may have major gene rearrangements that can only be identified with quantitative Southern blotting or fluorescence *in situ* hybridization analysis. |
| **Effect of mutation** | *VHL* is a tumor suppressor gene. In a patient carrying one mutant allele, somatic loss of the second allele results in tumor formation. |
| **Diagnosis** | The diagnosis requires a high index of suspicion, and can be made when the classical combination of clinical findings is observed. The specific diagnosis of VHL is confirmed by finding a mutation or deletion in one allele of the *VHL* gene. |
| **Counseling issues** | Early diagnosis of renal cell carcinoma, pheochromocytoma hemangioblastomas, and retinal angiomas can have a positive impact on the overall prognosis. Therefore, all family members carrying the mutation should be followed closely for the various tumors. |
| **References** | Richards FM. Molecular pathology of von Hippel-Lindau disease and the VHL tumour suppressor gene. *Exp Rev Mol Med* 2001. Available from URL: http://www-ermm.cbcu.cam.ac.uk/01002654h.htm. |

# Neurofibromatosis, Type 1

(also known as: NF1; von Recklinghausen disease)

| | |
|---|---|
| **MIM** | 162200 |

**Clinical features**    Clinical features include cutaneous neurofibromata, café-au-lait spots (see **Figure 5**), axillary freckling, optic glioma, meningioma, hypothalamic tumor, duodenal carcinoid, parathyroid adenoma, pheochromocytoma, renal artery stenosis, and hypertension. Precocious puberty may occur, particularly in patients with tumors of the optic chiasm. Central nervous system tumors occur in about 45% of patients. Secondary malignancies have been reported, with malignant degeneration of neurofibromas in 3%–15% of patients. Clinical symptoms may be secondary to structural changes due to lesions, such as stenosis of the bile duct, cerebral aqueduct, or renal artery.

**Primary tissue or**    Adrenal medulla.
**gland affected**

**Other organs, tissues,**    Parathyroid.
**or glands affected**

**Age of onset**    Symptoms usually lead to diagnosis in the first or second decade of life.

**Epidemiology**    The worldwide incidence of disease is estimated to be approximately 1:3500. Around half of the patients have familial disease, whereas *de novo* mutations are found in the remainder. These *de novo* mutations are more common on the paternal allele, and may be associated with increased paternal age. A higher incidence of disease, around 1:1000, has been reported in Israelis, particularly those of North African and Asian origin.

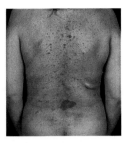

**Figure 5.** The skin-colored and pink-tan soft papules and nodules on the back are neurofibromata. A large café-au-lait spot is seen on the lower back. The large, soft, ill-defined, subcutaneous nodules on the right lower back and on the right posterior axillary line are plexiform neuromas. Reproduced with permission from the McGraw-Hill Companies (Fitzpatrick TB, Johnson RA, Wolff K et al., editors. *Color Atlas and Synopsis of Clinical Dermatology*, 3rd edn. New York: Mcgraw-Hill, 1997:461).

| | |
|---|---|
| **Inheritance** | Autosomal dominant. |
| **Chromosomal location** | 17q11.2 |
| **Genes** | *NF1* (neurofibromin). |
| **Mutational spectrum** | About 50% of cases have new mutations. More than 70 different mutations have been identified in the *NF1* gene, which spans at least 59 exons. |
| **Effect of mutation** | *NF1* appears to be a tumor suppressor gene, interacting with the protein product of the *RAS* proto-oncogene. Lesions are thought to be the result of mutations on both alleles. One is inherited and the second somatic, occurring on the normal allele. |
| **Diagnosis** | The clinical diagnosis is made on observing cutaneous neurofibromas, café-au-lait spots, and axillary freckling. Definitive diagnosis is based on finding mutations in the *NF1* gene. To make the diagnosis in the absence of genetic proof, at least six café-au-lait spots, each >1.5 cm in diameter, must be documented. |

| Counseling issues | About 50% of cases have new mutations; in these, there is essentially no genetic risk to siblings. |
| --- | --- |
| Notes | A similar syndrome, characterized by bilateral acoustic neuroma and meningioma with few cutaneous neurofibromas, is known as NF2 (MIM 101000). |
| References | MacCollin M, Kwiatkowski D. Molecular genetic aspects of the phakomatoses: Tuberous sclerosis complex and neurofibromatosis 1. *Curr Opin Neurol* 2001;14:163–9 (Review). |

# Familial Pheochromocytoma

(also known as: familial paraganglioma syndrome; sporadic pheochromocytoma)

**MIM** 171300

**Clinical features** Pheochromocytoma typically presents with episodic hypertension, associated with palpitations, a 'feeling of impending doom', and other symptoms of adrenergic overactivity. Many patients have persistent hypertension that may be resistant to conventional therapy. It has long been recognized that pheochromocytomas can appear as part of four well-defined clinical syndromes: von Hippel-Lindau syndrome (MIM 193300), multiple endocrine neoplasia (MEN)2a (MIM 171400), MEN2b (MIM 162300), and neurofibromatosis type 1 (MIM 162200). These syndromes are discussed in separate entries. In all four syndromes, pheochromocytoma may be the only finding at presentation, making identification of the specific syndrome very difficult. The need for genetic studies in patients with clinically sporadic disease is controversial and is discussed below.

**Primary tissue or gland affected** Adrenal medulla.

**Other organs, tissues, or glands affected** Paraganglioma.

**Age of onset** Sporadic and familial pheochromocytoma have been diagnosed at all ages between 4 years and >80 years old. Although tumors associated with germline mutations tend to appear earlier, there is considerable overlap, and age of diagnosis alone is of little help in identifying patients with familial disease.

**Epidemiology** Pheochromocytomas are rare and are found in only about 0.1% of patients presenting with hypertension.

| | |
|---|---|
| **Inheritance** | Autosomal dominant. |
| **Chromosomal location** | 11q23, 1p |
| **Genes** | *SDHB* (succinate dehydrogenase complex, subunit B), *SDHD* (succinate dehydrogenase complex, subunit D). |
| **Mutational spectrum** | Missense, nonsense, and splice-site deletions have been reported. Microinsertions and deletions, some of which cause frame shifts, have also been found. |
| **Effect of mutation** | The mutations prevent the production of functional protein. The mechanism by which mutations in these particular genes predispose to pheochromocytoma and paraganglioma is unknown. Recently, mutations in two subunits of the succinate dehydrogenase complex (SDHB and SDHD) have been identified in patients with familial paragangliomas of the neck (glomus tumors). Although none of the original families described had pheochromocytoma, subsequent study of families with familial pheochromocytomas revealed some with mutations in these genes. Thus, mutations in any of five genes (*SDHB*, *SDHD*, *RET*, *VHL*, and *NF1*) can cause familial pheochromocytoma. |
| **Diagnosis** | Biochemical diagnosis of pheochromocytoma is made by finding elevated levels of catecholamines or catecholamine metabolites in the blood or urine. Dynamic testing using clonidine or regitine may aid in the diagnosis of some patients. The tumor can be localized by CT or magnetic resonance imaging. MIBG scanning may be helpful for identifying extra-adrenal tumors and metastases. At the time of initial diagnosis, clinical tests should be performed to identify lesions associated with any of the syndromes mentioned above. If syndromic disease is suspected, the appropriate genetic tests should be performed to confirm the diagnosis and permit family screening. A family history should be obtained to identify familial disease. However, the majority of patients have negative family histories and no clinical evidence of syndromic disease. The need for genetic |

screening in these patients is still controversial. In a recent series of 271 patients with nonsyndromic pheochromocytomas and without a family history of the disease, 24% had mutations in one of the four genes studied (*RET*, *VHL*, *SDHB*, and *SDHD*). Young-onset, multifocal, and extra-adrenal tumors were significantly associated with the presence of a germline mutation. Clinical presentation before the age of 20 years was associated with mutations in about 60% of cases. However, in patients who presented after the age of 40 years, germline mutations were still identified in 8%. No mutations were identified in patients who presented after the age of 60 years. Eighty-four percent of multifocal tumors were hereditary, as were approximately 50% of extra-adrenal tumors. The latter were associated with mutations in *SDHB*, *SDHD*, or, less frequently, *VHL*. No *RET* mutations were identified in patients with extra-adrenal disease.

**Counseling issues**

The clinical importance of identifying mutations in any of these genes (*SDHB*, *SDHD*, *VHL*, or *RET*) cannot be overemphasized. Screening for mutations in *NF1* is much less rewarding due to the rarity of the syndrome and the size of the gene, making whole-gene mutation analysis prohibitively expensive and time-consuming. All of the syndromes are associated with potentially life-threatening disease. Diagnosis based on the presence of clinical symptoms is difficult and usually delayed. Thus, by the time clinical diagnosis is made, tumors may be inoperable, or may have caused irreversible damage. Thus, family screening and early diagnosis are critical to improve the long-term prognosis. However, the incidence of familial disease in patients with apparently sporadic pheochromocytoma is still controversial, and the cost-effectiveness of screening all potential genes for novel mutations is still not proven. A recent study of a very large population of sporadic cases strongly suggested that screening for mutations in four genes (*SDHB*, *SDHD*, *VHL*, or *RET*) is indeed very cost-effective, especially if the proband has multifocal or extra-adrenal disease, or if the tumor is diagnosed before the age of 20 years. For patients diagnosed after the age of 40 years with single, adrenal pheochromocytomas and no family history, the cost-effectiveness of genetic screening is more difficult to prove, and additional, confirmatory studies are needed.

**Notes**

The discovery of familial pheochromocytomas in patients with mutations in *SDHB* and *SDHD* has greatly increased the yield of genetic screening of patients with apparently sporadic disease. Although the incidence of pheochromocytoma in patients with apparently sporadic glomus tumors has not been defined, it seems prudent to screen these patients for occult pheochromocytoma, prior to surgical intervention.

**References**

Bar M, Friedman E, Jakobovitz O et al. Sporadic phaeochromocytomas are rarely associated with germline mutations in the von Hippel-Lindau and RET genes. *Clin Endocrinol (Oxf)* 1997;47:707–12.

Brauch H, Hoeppner W, Jahnig H et al. Sporadic pheochromocytomas are rarely associated with germline mutations in the vhl tumor suppressor gene or the ret protooncogene. *J Clin Endocrinol Metab* 1997;82:4101–4.

Neumann HP, Bausch B, McWhinney, SR et al. Germ-line mutations in nonsyndromic pheochromocytoma. *N Engl J Med* 2002;346:1459–66.

# 7. Genetic Defects of the Reproductive Glands

# Genetic defects of the reproductive glands

This chapter contains monogenic diseases that are primarily characterized by gonadal abnormalities and may present with varying degrees of genital ambiguity, sex reversal (genetic males with a female phenotype), or hypogonadism. In addition, we have also included: Klinefelter's syndrome, which is a chromosomal abnormality, and polycystic ovary disease, which appears to have a complex genetic etiology in most cases.

Many of these syndromes are caused by mutations in transcription factors that are responsible for the regulation of normal gonadogenesis. These factors and their primary functions are shown in **Figure 1**. Others are caused by mutations in enzymes responsible for androgen and estrogen synthesis, and are illustrated in **Figure 2**.

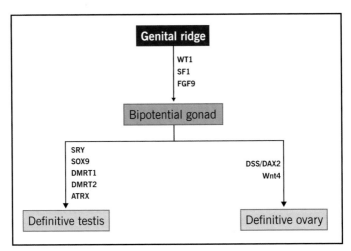

**Figure 1.** The major transcription factors required for gonadogenesis. Mutations in some of these cause syndromes associated with primary endocrine defects that are discussed in this book, whereas others are associated with syndromes in which the primary phenotype is not endocrine. The Wilm's tumor gene (WT1) is involved in Denys–Drash, Frasier, and WAGR sydromes; steroidogenic factor 1 (SF1, FTZ-F1) is involved in syndrome of XY sex reversal and adrenal failure; homozygous mutations of fibroblast growth factor 9 (FGF9) are associated with pulmonary hypoplasia and XY sex reversal; the testis-determining factor on the Y chromosome, SRY, is responsible for some cases of pure gonadal dysgenesis; SOX9 is a SRY-homeobox-related gene on 17q24.3-q25.1, mutations in which cause campomelic dysplasia (MIM 114290) and skeletal dysplasia with gonadal and genital malformations; DMRT1 and DMRT2 are doublesex and MAB-3 related transcription factors that are candidate genes for the 9p sex-reversal syndromes; ATR-X is responsible for α thalassemia, mental retardation, and X-linked syndrome with multiple anomalies, including 46XY gonadal dysgenesis; dosage-sensitive sex reversal (DSS)/DAX-1 may inhibit testicular differentiation and be responsible for X-linked adrenal hypoplasia congenita; Wnt-4 located on chromosome 1p35 appears necessary for Müllerian duct and normal ovarian development. Reproduced with permission from Blackwell Scientific Publications (*Clin Endocrinol* 2002;56:1–18).

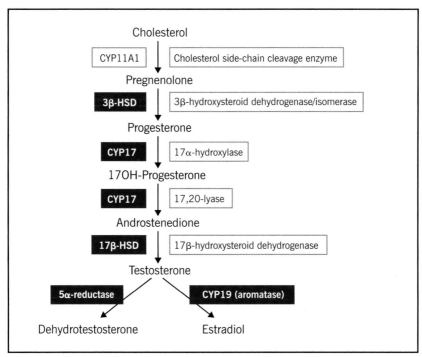

**Figure 2.** The major steps in sex hormone biosynthesis. Enzymes that are associated with specific syndromes described in this book are highlighted in purple: the first two, 3β-hydroxysteroid dehydrogenase (HSD) (MIM 201810) and CYP17 (MIM 202110) are described in Chapter 6, since the primary clinical presentation relates to defects in adrenal steroidogenesis. The rest, 17β-HSD (MIM 264300), 5α reductase (MIM 264600), and CYP19 (MIM 107910), are described in this chapter.

Diseases characterized by genital virilization due to other steroid biosynthetic defects are discussed in Chapter 6. Multiple pituitary hormone deficiency syndromes and Turner's syndrome, which typically present initially with short stature, are described in Chapter 2. The iron overload syndromes, which can present with hypogonadism later in life, are discussed in Chapter 1.

# Persistent Müllerian Duct Syndrome

(also known as: PMDS; hernia uteri inguinale; female genital ducts in an otherwise normal male)

| | |
|---|---|
| **MIM** | 261550 |

**Clinical features**

Affected patients are 46,XY males with normal testes and normal male external genitalia, in whom Fallopian tubes and a uterus are discovered during surgery, usually for inguinal hernia repair (see **Figure 3**). Most often, the hernia contains a partially or fully descended testis together with the ipsilateral Fallopian tube and uterus. Sometimes, the contralateral testis and tube may be present in the same hernia sac; this condition is known as transverse testicular ectopia. Less commonly, PMDS presents as bilateral cryptorchidism, with the uterus fixed in the pelvis and both testes present in the broad ligament.

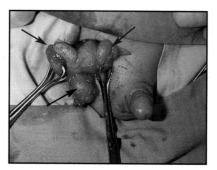

**Figure 3.** This picture was taken during surgery for an incarcerated inguinal hernia in a 12-day-old boy with persistent Müllerian duct syndrome (due to anti-Müllerian hormone deficiency). The smooth, oval-shaped structures above the retractors are the right and left testes, respectively (purple arrows). The smooth, 'domed' structure between them is the uterus. The structure below the uterus is the round ligament (black arrow). Courtesy of Professor Omri Lernau, Shaare Zedek Medical Center, Jerusalem, Israel.

**Primary tissue or gland affected**

Female reproductive tract.

**Other tissues affected**

Testes

| Age of onset | PMDS is discovered during inguinal hernia repair, orchiopexy, or other abdominal surgery, usually in infancy or childhood. |
|---|---|
| Epidemiology | PMDS is a rare condition, of which approximately 150 cases have been described. Mutations in the gene for anti-Müllerian hormone (*AMH*) are found in 45% of PMDS patients, and for AMH type 2 receptor (*AMHR2*) in nearly 40%. No mutations are found in the remaining 15% of affected families. *AMH* mutations are most common in patients from Mediterranean and Arab countries, while *AMHR2* mutations are more frequently found in those from France and northern Europe. |
| Inheritance | Sex limited, autosomal recessive. |
| Chromosomal location | 19p13.3-p13.2, 12q13 |
| Genes | *AMH* (anti-Müllerian hormone), *AMHR2* (anti-Müllerian hormone receptor type 2). |
| Mutational spectrum | Splicing, missense, nonsense, and deletion mutations may be present throughout the *AMH* gene, but occur most commonly in exon 1 and in the C-terminal half of exon 5. Deletion, missense, and nonsense mutations are identified throughout the *AMHR2* gene; a 27-bp deletion in exon 10 is the most common. |
| Effect of mutation | Binding of AMH, a dimeric glycoprotein member of the transforming growth factor-$\beta$ family, to the type 2 receptor activates a signal transducer (type 1 receptor), which then phosphorylates downstream signaling molecules. Failure of the fetal Sertoli cells to secrete AMH, or resistance of the target organs at around 8 weeks gestation (when the Müllerian ducts are normally responsive to AMH) results in the persistence of Müllerian duct derivatives in otherwise normal males. |

**Diagnosis**

Confirmation of PMDS requires the typical clinical features, a normal male karyotype, and demonstration of normal testicular histology (no evidence of ovotestes). A testicular biopsy is usually performed when Müllerian structures are discovered during surgery for inguinal hernia or orchiopexy. Ambiguity of the external genitalia is not consistent with PMDS. There is no difference in clinical presentation between patients with AMH deficiency and those with type 2 receptor defects. In normal males, serum AMH levels decline to low or undetectable levels by puberty. A normal AMH level in a prepubertal boy with PMDS indicates a receptor defect.

**Counseling issues**

Women who are homozygous for *AMH* gene deletions have normal internal and external genitalia, and normal fertility. Infertility is common in affected men, and is most likely secondary to torsion of the testes, late age at orchiopexy, and technical difficulties in freeing the vas deferens from entrapment by Müllerian structures. The risk of testicular neoplasia in men with PMDS, including carcinoma *in situ*, seminomatous, and other germ cell tumors, may be as high as 15%.

**Notes**

Because excision of all Müllerian remnants may increase the risk of surgical damage to the internal spermatic vessels, it has been suggested that Müllerian structures should be left *in situ*.

**References**

Belville C, Josso N, Picard JY. Persistence of Müllerian derivatives in males. *Am J Med Genet* 1999;89:218–23 (Review).

Vandersteen DR, Chaumeton AK, Ireland K et al. Surgical management of persistent Müllerian duct syndrome. *Urology* 1997;49:941–5.

# Familial Male Precocious Puberty

(also known as: FMPP; testotoxicosis; male-limited gonadotropin-independent sexual precocity)

| | |
|---|---|
| **MIM** | 176410 |
| **Clinical features** | The presenting signs of FMPP are testicular enlargement, penile growth, and the appearance of pubic hair in early childhood. Other features of precocious puberty include rapid linear growth, accelerated skeletal maturation, and prominent muscular development. |
| **Primary tissue or gland affected** | Testes. |
| **Age of onset** | The first signs of precocious puberty usually appear in early childhood, generally by the age of 4 years, but some cases of penile and testicular enlargement in newborn infants have been reported. |
| **Epidemiology** | Fifteen activating mutations of the luteinizing hormone (LH) receptor gene have been described in more than 60 patients with FMPP. |
| **Inheritance** | Male limited, autosomal dominant. Some sporadic cases have been described. |
| **Chromosomal location** | 2p21 |
| **Genes** | *LHR* (LH receptor). |
| **Mutational spectrum** | Missense mutations in exon 11 of *LHR* account for all cases of FMPP. Exon 11 encodes the carboxy-terminal half of the LH receptor, including all seven transmembrane helices. Most of the gain of function mutations are clustered within the sixth transmembrane helix and third intracellular loop. |
| **Effect of mutation** | Mutations result in constitutive activation of the LH receptor, leading to markedly increased production of cAMP. |

**Diagnosis**

Testosterone levels are elevated to the pubertal or adult male range, despite prepubertal levels of LH. Gonadotropin-releasing hormone (GnRH)-stimulated gonadotropin levels are low, consistent with a prepubertal response, and pulsatile LH secretion is not observed. In late childhood, secondary, 'true' central gondadotropin-dependent precocious puberty may develop. The diagnosis of underlying testotoxicosis may be difficult to confirm when combined central and peripheral precocious puberty coexist. Disorders that need to be excluded in suspected FMPP patients include human chorionic gonadotropin-producing tumors, testosterone-producing testicular or adrenal tumors, McCune–Albright syndrome, and congenital adrenal hyperplasia.

**Counseling issues**

Spermatogenesis and fertility are usually normal in affected men. Female carriers are unaffected.

**Notes**

GnRH analogs are ineffective in treating FMPP, except as ancillary treatment when secondary central precocious puberty develops. Medroxyprogesterone, ketoconazole, and testolactone have been used with variable degrees of success. Newer antiandrogens that may be useful include flutamide and letrozole.

**References**

Latronico AC, Segaloff DL. Naturally occurring mutations of the luteinizing-hormone receptor: Lessons learned about reproductive physiology and G protein-coupled receptors. *Am J Hum Genet* 1999;65:949–58 (Review).

Richter-Unruh A, Wessels HT, Menken U et al. Male LH-independent sexual precocity in a 3.5-year-old boy caused by a somatic activating mutation of the LH receptor in a Leydig cell tumor. *J Clin Endocrinol Metab* 2002;87:1052–6.

# Androgen Insensitivity Syndrome

(also known as: AIS; androgen resistance; testicular feminization; Reifenstein syndrome;
X-linked male pseudohermaphroditism)

| | |
|---|---|
| **MIM** | 300068, 312300 |
| **Clinical features** | Disorders of the androgen receptor are associated with a remarkably wide range of clinical phenotypes. Affected patients are 46,XY males, with bilateral testes and unimpaired testosterone synthesis. Prior to puberty, the testes are histologically normal. After puberty, the seminiferous tubules are often small, spermatogenesis is absent due to maturational arrest, and the Leydig cells tend to be hyperplastic. Five clinical phenotypes have been described: complete AIS, incomplete AIS, Reifenstein syndrome, infertile male syndrome, and undervirilized fertile male syndrome. The phenotype in complete AIS is that of a normal female, except for sparse or absent axillary and pubic hair, an absent or short and blind vagina, and, except for testes, absent internal genital structures. The phenotype of incomplete AIS is predominantly female, but with normal axillary and pubic hair, variable degrees of clitoromegaly and labial fusion, and the presence of Wolffian duct structures (epididymis, vas deferens, seminal vesicles, and ejaculatory ducts), though these are less developed than in normal men. The Reifenstein syndrome phenotype is predominantly male, with variable degrees of undervirilization. These range from a microphallus with a normal penile urethra, to the more common appearance of penoscrotal hypospadias with a bifid scrotum, and the extreme case of complete failure of scrotal fusion with a pseudovagina. Testes may be either undescended or scrotal, but are usually small, and spermatogenesis is impaired. Gynecomastia is common at puberty, axillary and pubic hair are normal, but hair on the face and chest is sparse. In the infertile male syndrome, the general appearance, including facial and body hair, is that of a normal male. Male external and internal genitalia development are normal and the testes are usually fully descended, but variable |

defects in germinal cell structure or spermatogenesis result in infertility. The undervirilized fertile male syndrome describes a small group of men with normal internal genitalia, scrotal testes, and a small penis. Sperm density is normal, but ejaculate volume may be decreased. Facial and body hair are sparse. Some men with this phenotype are fertile.

| | |
|---|---|
| **Primary tissue or gland affected** | Gonads. |
| **Other organs, tissues, or glands affected** | Androgen-sensitive tissues, such as the accessory organs of male reproduction. In the testes, androgen receptors are present both in Sertoli and Leydig cells. Tissues such as skeletal muscle, the heart, and placenta have small numbers of androgen receptors. |
| **Age of onset** | The age at diagnosis ranges from infancy to adulthood, depending on the specific phenotype. |
| **Epidemiology** | The incidence of complete AIS may be as high as 1:20 000, and, according to some series, it is the third most frequent cause of primary amenorrhea, after gonadal dysgenesis and congenital absence of the vagina. The syndrome of incomplete AIS (in the narrow sense of patients with a female phenotype resembling complete AIS, but with normal body hair and partial virilization of the external genitalia) is about one tenth as common as the complete form. The frequency of Reifenstein syndrome is difficult to assess, but may be as common as complete AIS. Partial AIS (infertile male syndrome) was found in 10% of men evaluated for idiopathic azoospermia or severe oligospermia. The frequency of undervirilized, fertile male syndrome is unknown. |
| **Inheritance** | X-linked. |
| **Chromosomal location** | Xq11-q12 |

| Genes | AR (androgen receptor gene) |

**Genes**      AR (androgen receptor gene)

**Mutational spectrum**      Deletions, splicing abnormalities, premature termination codons, and amino acid substitutions in either the hormone-binding or DNA-binding domains of the AR gene have been observed.

**Effect of mutation**      Gene deletions, splicing abnormalities, and premature termination codons result in the synthesis of mutant androgen receptors that are unable to bind ligands (see **Figure 4** for the mechanisms of androgen action). These mutations, which have been localized to all eight exons of the androgen receptor, result in severe androgen resistance, and are associated with the syndrome of complete testicular feminization. Amino acid substitutions in either the hormone-binding or DNA-binding domains have been found in families with androgen receptor function ranging from absent or near-absent binding, qualitatively abnormal binding, decreased numbers of qualitatively normal receptors, and, in some families, normal receptor binding despite clinical evidence of androgen resistance. Amino acid substitutions are responsible for the wide range of phenotypes seen in complete and incomplete forms of AIS.

**Diagnosis**      The diagnosis of AIS needs to be considered when evaluating the following people: infants with ambiguous genitalia; females with inguinal masses (1%–2% of girls with inguinal hernia may have AIS); adolescent girls with primary amenorrhea or signs of virilization and clitoromegaly; adolescent boys who have failed to undergo normal puberty or with persistent gynecomastia; and adult men who appear undervirilized or who are infertile. Depending on the clinical presentation, differential diagnoses include congenital absence of the vagina (Müllerian agenesis), mixed gonadal dysgenesis, defects in testosterone synthesis, gonadotropin deficiency, and 5α-reductase deficiency. The diagnosis of AIS is supported by a 46,XY karyotype, the identification of testes in the abdomen or inguinal region by ultrasound, other imaging techniques, or laparoscopy, and serum testosterone levels in the range for normal men. A normal ratio

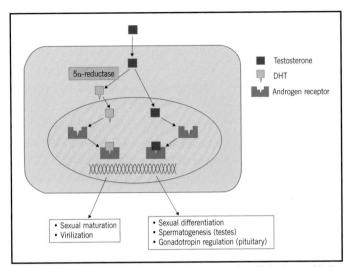

**Figure 4.** The molecular mechanism of androgen action. Testosterone binds to the androgen receptor either directly or after intracellular conversion to dihydrotestosterone (DHT). DHT binds more strongly to the receptor than testosterone, and the DHT–receptor complex is a stronger regulator of gene expression. Thus, the two receptor complexes regulate transcriptional activity somewhat differently. The major actions of testosterone are shown on the right, whereas the major actions of DHT are shown on the left.

of basal or human chorionic gonadotropin-stimulated testosterone to dihydrotestosterone (DHT) excludes 5α-reductase deficiency. In newborn males with a micropenis, small doses of intramuscular testosterone enanthate may be given to increase penile growth. Because of difficulties in studying androgen receptor binding in genital skin fibroblasts, and because the large number of mutations makes DNA screening impractical, the determination of specific androgen receptor defects is generally recommended only when a clinical diagnosis cannot be made using information from family history and hormonal studies, and when the specific diagnosis will directly affect therapeutic decisions.

**Counseling issues**

Decisions regarding gender assignment have traditionally been made in infancy, based on anatomic and surgical considerations, since in most cases of AIS the potential for fertility is low or nonexistent. The medical team, including an endocrinologist, urologist or gynecologist, and psychologist with experience in problems of sexual differentiation, should explain the nature of the specific problem to the parents. In the past, the final decision regarding gender assignment was delegated to the parents. Recently, it has been proposed that the final decision should be postponed until the patient is old enough to decide whether they wish to undergo surgery for gender reassignment. Complete AIS patients have a high risk of developing testicular tumors, and orchiectomy is therefore usually performed after pubertal (breast) development is complete.

**Notes**

In some families, the same mutation may result in variable phenotypic expression of AIS. Those mutations accounting for variable phenotypes have been estimated to represent about 25% of the amino acid replacements, and occur within the DNA-binding and hormone-binding domains of the androgen receptor. The different phenotypes may be due to variable timing of receptor expression, differences in testosterone synthesis or metabolism, or influences of transcription factors or other genes, which may accentuate or ameliorate the functional effects of the androgen receptor mutation. Because of the decreased mass of urogenital tissues, where DHT is usually produced, some patients with Reifenstein syndrome may have an elevated testosterone to DHT ratio, giving the mistaken impression of $5\alpha$-reductase deficiency.

**References**

Griffin JE. Androgen resistance – the clinical and molecular spectrum. *N Engl J Med* 1992;326:611–8 (Review).

McPhaul MJ, Griffin JE. Male pseudohermaphroditism caused by mutations of the human androgen receptor. *J Clin Endocrinol Metab* 1999;84:3435–41 (Review).

# Male Pseudohermaphroditism and Renal Disease

(includes: male pseudohermaphroditism due to WT1 mutations; Denys–Drash syndrome [DDS]; Frasier syndrome [FS]; Wilms' tumor, aniridia [WAGR]; genitourinary anomalies; mental retardation syndrome)

| | |
|---|---|
| **MIM** | 194080 (DDS), 136680 (FS), 194072 (WAGR) |
| **Clinical features** | Children with DDS have a 46,XY karyotype, ambiguous or female genitalia, streak gonads, and nephropathy (diffuse mesangial sclerosis), with Wilms' tumor appearing in >50% of patients. Like DDS, FS is characterized by female genitalia in a 46,XY male with rudimentary gonads. In contrast to DDS, renal disease in FS is focal and segmental glomerulosclerosis, and these patients do not develop Wilms' tumors. FS patients often develop gonadoblastomas. Females (46,XX) with Wilms' tumor gene mutations have renal disease with normal external genitalia. Gonadal anomalies include ovotestes or immature, streak gonads. Recently, severe obesity has been described in some patients with WAGR, in addition to predisposition to Wilms' tumor, aniridia, genitourinary anomalies, and mental retardation. |
| **Primary tissue or gland affected** | Kidneys, fetal gonads. |
| **Other organs, tissues, or glands affected** | Internal and external genitalia; eye and central nervous system (WAGR). |
| **Age of onset** | Deficient function of the fetal testes results in the genital anomalies noted at birth. Proteinuria appears in infancy, and end-stage renal failure occurs by the age of 3 years in DDS. Wilms' tumor is diagnosed in children before the age of 2 years in DDS, but may appear later in WAGR syndrome. |
| **Epidemiology** | About 150 cases of DDS have been reported; FS is less frequent. In a study of 2000 patients with Wilms' tumor, 22 had DDS and 46 had WAGR syndrome. |

| Inheritance | Autosomal dominant. |
|---|---|
| Chromosomal location | 11p13 |
| Genes | *WT1* (Wilms' tumor gene). |
| Mutational spectrum | *WT1* is a 10-exon gene that encodes a transcription factor with tumor-suppressing activity. Point mutations in exons 6–9 or introns 6 and 9 are responsible for DDS. The most common DDS-causing mutation is a substitution of 394Arg with Trp. Mutations have been identified in 90% of patients with DDS. FS results from mutations at the donor splice site in intron 9 of *WT1*. WAGR appears to be caused by a contiguous gene deletion at 11p13 (microdeletion). Loss of an allele of *PAX6* results in aniridia, and loss of (or mutations in) *WT1* is associated with the observed genitourinary anomalies. |
| Effect of mutation | In DDS, mutations produce a truncated protein, loss of ligand sites, and defective DNA binding, and exert a dominant negative effect on the native protein. In FS, mutations alter the ratio of WT1 protein isoforms; wild-type protein contains a 3-amino acid segment ('KTS'), which is lost due to abnormal splicing. |
| Diagnosis | The presence of renal disease in an infant with male pseudohermaphroditism strongly suggests the diagnosis of DDS or FS. Karyotyping and testing for *WT1* mutations have been recommended in infants with early-onset nephrotic syndrome. Cytogenetic studies that identify a microdeletion at 11p13 will confirm the diagnosis of WAGR. |
| Counseling issues | Infants with ambiguous genitalia should be screened for renal disease, and followed for its development. In phenotypically female infants with early-onset nephropathy, the possibility of male pseudohermaphroditism needs to be considered. Wilms' tumor develops in >50% of infants with DDS, while FS infants are at risk for gonadoblastoma. |

**Notes**

There appears to be considerable overlap between DDS and FS phenotypes. Recent reports suggest that it may be more appropriate to consider these two syndromes within the spectrum of *WT1* mutation disorders, rather than as separate entities.

**References**

McTaggart SJ, Algar E, Chow CW et al. Clinical spectrum of Denys–Drash and Frasier syndrome. *Pediatr Nephrol* 2001;16:335–9.

Neri G, Opitz J. Syndromal (and nonsyndromal) forms of male pseudohermaphroditism. *Am J Med Genet* 1999;89:201–9 (Review).

# XY Pure Gonadal Dysgenesis

(also known as: Swyer syndrome; XY sex reversal)

| | |
|---|---|
| **MIM** | 306100 |
| **Clinical features** | Until puberty, XY females have an unambiguously female phenotype, including a vagina, uterus, and Fallopian tubes. Patients seek medical evaluation at puberty, when breast development and menses fail to occur. Gonadotropin levels are elevated, but somatic features of Turner's syndrome are absent. Gonadal tissue consists of undifferentiated streaks, usually with an ovarian-like stroma, and an increased tendency to develop gonadoblastomas or dysgerminomas. |
| **Primary tissue or gland affected** | Fetal primitive gonad. |
| **Age of onset** | Early in embryonic development. |
| **Epidemiology** | This condition is very rare. |
| **Inheritance** | Most cases represent *de novo* mutations, but some familial cases with an X-linked or sex-limited autosomal dominant pattern of transmission have been reported. |
| **Chromosomal location** | Yp11.3, Xp22.11-p21.2. |
| **Genes** | *SRY* (testis-determining factor; also known as *TDF*). However, mutations in *SRY* are found in only 15%–20% of patients with XY pure gonadal dysgenesis. |
| **Mutational spectrum** | Substitutions, frame-shift mutations, and deletions have been reported. SRY is located near the pseudoautosomal region of the Y chromosome. Aberrant recombinations sometimes result in the transfer of *SRY* to the X chromosome. |

*Genetic Defects of the Reproductive Glands*

**Effect of mutation**    Most mutations occur in the high mobility group domain, and appear to either prevent DNA binding or change the DNA-bending angle. As a result of mutant *SRY*, the genital ridge is unable to differentiate into a normal testis.

**Diagnosis**    This condition should be distinguished from XY gonadal agenesis, in which functional testicular tissue is present for long enough during fetal development to prevent development of Müllerian ducts. Patients with XY gonadal agenesis have no gonadal tissue or uterus, and usually no vagina.

**Counseling issues**    Because of the risk of neoplasia, streak gonads should be removed. Pregnancy may be possible using a donor embryo and assisted reproduction technology.

**Notes**    Translocation of the *SRY*-containing part of the Y chromosome to an X chromosome results in a male phenotype with an XX genotype. Another defect associated with XY sex reversal involves a serine-threonine protein kinase gene (*PRKY*) on the Y chromosome, which undergoes recombination to form *PRKX*. XY sex reversal can also be seen with autosomal defects, including terminal deletion of 10p, monosomy of 9p, 9p deletion, and duplication of 1p.

**References**    McElreavey K, Fellous M. Sex determination and the Y chromosome. *Am J Med Genet* 1999;89:176–85 (Review).

# 17β-Hydroxysteroid Dehydrogenase Type 3 Deficiency

(also known as: 17β-ketosteroid reductase deficiency; male pseudohermaphroditism [with gynecomastia] due to 17β-hydroxysteroid dehydrogenase deficiency)

| | |
|---|---|
| **MIM** | 264300 |
| **Clinical features** | In affected XY male infants, the external genitalia are predominantly (and often unambiguously) female. The discovery of inguinal masses or an abnormal genital appearance in infancy or childhood are the most common reasons for seeking medical care. The characteristic features of 17β-hydroxysteroid dehydrogenase (HSD)3 deficiency include male pseudohermaphroditism with no abnormalities in adrenal steroid production, absence of Müllerian duct derivatives, and normal Wolffian structures. The testes and epididymis may be present in the inguinal canal. At puberty, patients undergo virilization, including phallic enlargement, beard and body hair growth, and deepening of the voice. Activity of the more widely expressed type 1 isozyme of 17β-HSD accounts for virilization at puberty. As described for 5α-reductase deficiency, some patients with 17β-HSD3 deficiency who were initially raised as girls change their gender role from female to male at puberty. Variable degrees of gynecomastia appear in approximately 50% of patients at adolescence. |
| **Primary tissue or gland affected** | Testes. |
| **Other organs, tissues, or glands affected** | External genitalia, Wolffian duct derivatives (epididymis, vas deferens, seminal vesicles, and ejaculatory ducts), breast tissue (gynecomastia). |
| **Age of onset** | Affected males fail to virilize external genitalia *in utero*. Virilization and gynecomastia appear at puberty. |
| **Epidemiology** | The incidence of 17β-HSD3 deficiency varies with different ethnic groups. In the Netherlands, 17β-HSD3 deficiency occurs in |

1:147 000 neonates, an incidence comparable with that of androgen insensitivity syndrome. Among the highly inbred population of Arabs in Gaza, the incidence is approximately 1:200–300.

| | |
|---|---|
| **Inheritance** | Autosomal recessive. |
| **Chromosomal location** | 9q22 |
| **Genes** | *HSD17B3* (17β-hydroxysteroid dehydrogenase type 3). |
| **Mutational spectrum** | Frame shifts, splicing defects, and missense mutations have been reported. |
| **Effect of mutation** | 17β-HSD converts androstenedione to testosterone. The type 3 isozyme is a microsomal enzyme expressed in the testis. Missense mutations most often result in the total absence of 17β-HSD3 activity; partial activity was found with the Arg80Gln missense mutation. |
| **Diagnosis** | Patients with clinical features suggestive of 17β-HSD3 deficiency should undergo a human chorionic gonadotropin (hCG)-stimulation test to determine the testosterone:androstenedione (T:A) ratio, which is abnormally low in 17β-HSD3-deficient patients due to their inability to convert androstenedione to testosterone in the absence of normal enzyme activity. Evaluation of the T:A ratio without prior hCG stimulation may be misleading, since similar findings can be seen in other defects of testosterone synthesis and in Leydig cell hypoplasia. Demonstration of homozygous or compound heterozygous mutations confirms the diagnosis. Knowing the patient's ethnic background simplifies the search for mutations, which are closely associated with geographic origin. |

**Counseling issues**　Affected men are infertile. Possible explanations for the lack of spermatogenesis include cryptorchidism and low levels of intratesticular testosterone. Most newborn infants with this condition are raised as females. When a male gender role is chosen, testosterone injections to increase phallic size are most effective in infancy.

**Notes**　Some patient advocacy groups recommend the postponement of genital surgery until the child is sufficiently mature to chose his or her own gender role.

**References**　Boehmer AL, Brinkmann AO, Sandkuijl LA et al. 17Beta-hydroxysteroid dehydrogenase-3 deficiency: Diagnosis, phenotypic variability, population genetics, and worldwide distribution of ancient and de novo mutations. *J Clin Endocrinol Metab* 1999;84:4713–21.

# Idiopathic Hypogonadotropic Hypogonadism

(also known as: IHH; isolated gonadotropin deficiency [IGD]; Kallmann syndrome [IHH with anosmia or hyposmia]; congenital gonadotropin-releasing hormone [GnRH] deficiency)

| | |
|---|---|
| **MIM** | 146110 (IHH), 308700 (Kallmann syndrome), 138850 (GnRH receptor defects) |
| **Clinical features** | Manifestations of hypogonadism include the following: micropenis and cryptorchidism; delayed, absent, or arrested puberty, including amenorrhea; and eunuchoid body proportions (arm span exceeds height by >5 cm, and upper/lower body segment ratio is <1.0). Growth is normal during childhood, and pubic hair may develop as a result of adrenal androgen secretion. Testicular volume is usually very small and gynecomastia is not seen. Patients with congenital GnRH deficiency may have other congenital anomalies, such as midline cleft lip or palate, unilateral renal agenesis, and color blindness. Unilateral renal agenesis and 'mirror movements' (synkinesia) are markers for the X-linked form of Kallman syndrome. |
| **Primary tissue or gland affected** | Hypothalamus and anterior pituitary. |
| **Other organs, tissues, or glands affected** | Testes and ovaries. |
| **Age of onset** | The earliest presentation of IHH occurs in newborn boys with a micropenis. Delayed puberty and primary amenorrhea are the presenting features in adolescence. Other patients are diagnosed because of pubertal arrest, hypogonadism, or infertility as adults. |
| **Epidemiology** | The prevalence of Kallmann syndrome is 1:7500 men and 1:50 000 women. GnRH deficiency is the most common etiology; a defect in the GnRH receptor is present in only 5% of IHH cases. Approximately one third of IHH patients have a positive family history, of whom 34% show X-linked inheritance. |

| Inheritance | X-linked, autosomal dominant, autosomal recessive, sporadic. |
|---|---|
| Chromosomal location | Xp22.3 (KAL or KALIG1), 4q21.2 (GnRH receptor) |
| Genes | *KAL* or *KALIG1* (anosmin 1); *GnRHR* (GnRH receptor). The genes responsible for other autosomal forms of Kallmann syndrome have not yet been identified. |
| Mutational spectrum | Deletions, point mutations, and nonsense mutations within the *KAL* gene have been described. Most *KAL* mutations occur in the carboxy-terminal fibronectin-3 region of the protein anosmin. Point mutations in the *GnRHR* gene have been found in the amino terminus, first, second, and third extracellular loops, and transmembrane domains 3–7. |
| Effect of mutation | *KAL* mutations cause frame-shift and premature stop codons, producing an abnormal protein which disrupts normal GnRH neuronal growth and migration. *GnRHR* mutations are associated with loss of receptor function. Different *GnRHR* mutations that result in similar functional changes in the receptor may be associated with significant variations in the clinical phenotype. |
| Diagnosis | It is very difficult to distinguish IHH from constitutional delayed puberty. The laboratory criteria for IHH include prepubertal serum levels of testosterone in boys, and estradiol in girls. Luteinizing hormone (LH) and follicle-stimulating hormone (FSH) levels are in the low to normal range. Because of considerable overlap in gonadotropin responses, it is not usually possible to distinguish IHH from constitutional delayed puberty or functional hypogonadotropic hypogonadism by means of the GnRH stimulation test alone. Other abnormalities of pituitary function need to be excluded. Magnetic resonance imaging (MRI) may show aplasia or hypoplasia of the olfactory lobes. A CT scan or MRI must be performed in order to rule out tumors or other central nervous system (CNS) lesions. Formal smell testing should be performed to confirm or exclude |

anosmia/hyposmia. Hypogonadotropic hypogonadism may occur together with congenital adrenal hypoplasia due to mutations of the *DAX1* gene. Isolated LH deficiency in men ('fertile eunuch syndrome') may be caused by CNS tumors, or rarely by a mutation in the LH β-chain gene. Isolated FSH deficiency in women due to FSH β-chain mutations results in amenorrhea and variable degrees of delayed puberty.

**Counseling issues**

With appropriate hormonal replacement (gonadotropin or pulsatile GnRH), the prognosis for fertility is good.

**Notes**

Hypogonadotropic hypogonadism may also be seen in other conditions, including anorexia nervosa, amenorrhea associated with athletic training, Prader–Willi syndrome, Bardet–Biedl syndrome, X-linked cerebellar ataxia in males, and the multiple lentigines and basal cell nevus syndromes.

**References**

Achermann JC, Weiss J, Lee EJ et al. Inherited disorders of the gonadotropin hormones. *Mol Cell Endocrinol* 2001;179:89–96 (Review).

Beranova M, Oliveira LM, Bedecarrats GY et al. Prevalence, phenotypic spectrum, and modes of inheritance of gonadotropin-releasing hormone receptor mutations in idiopathic hypogonadotropic hypogonadism. *J Clin Endocrinol Metab* 2001;86:1580–8.

Seminara SB, Crowley WF Jr. Perspective: The importance of genetic defects in humans in elucidating the complexities of the hypothalamic–pituitary–gonadal axis. *Endocrinology* 2001;142:2173–7 (Review).

# Steroid 5α-Reductase Type 2 Deficiency

(also known as: male pseudohermaphroditism due to 5α-reductase deficiency [5aRD]; pseudovaginal perineoscrotal hypospadias)

| | |
|---|---|
| **MIM** | 264600 |
| **Clinical features** | The classical phenotype of 5aRD is a 46,XY male with testes, a male ejaculatory duct system which ends in a urogenital sinus (blind-end vagina), and a mainly female appearance of the external genitalia at birth. Testosterone levels are normal, while dihydrotesterone (DHT) is low, and the testosterone to DHT ratio is increased. Newborn males have ambiguous genitalia with a small penis, hypospadias, bifid scrotum, and a urogenital sinus that opens on the perineum. The testes may be scrotal or in the inguinal canal. Considerable phenotypic variability can be seen, ranging from a normal female appearance to an almost normal appearance for an infant male, except for hypospadias. The Wolffian ducts (vas deferens, epididymis, and seminal vesicles) are normal. A characteristic feature of 5aRD is virilization of the external genitalia at puberty. |
| **Primary tissue or gland affected** | External genitalia. |
| **Age of onset** | Insufficient production of DHT *in utero* results in variable degrees of undervirilization in the newborn male. In more severely affected males, in whom the phenotype may be unambiguously female, 5aRD may be diagnosed at adolescence during investigation of amenorrhea. |
| **Epidemiology** | This is a rare disorder: approximately 50 affected families have been reported. The largest number of cases are from the Dominican Republic, Brazil, Turkey, and Papua New Guinea. |
| **Inheritance** | Autosomal recessive. |

| | |
|---|---|
| **Chromosomal location** | 2p23 |
| **Genes** | *SRD5A2* (steroid 5α-reductase type 2). |
| **Mutational spectrum** | Gene deletions, point mutations, nonsense mutations, splicing defects, and missense mutations have been identified. Missense mutations are the most frequent. About 60% of affected patients are compound heterozygotes, while the rest are homozygous. Mutations are found throughout all five exons of the *SRD5A2* gene. |
| **Effect of mutation** | No detectable enzyme activity could be found in half of the cases described. Decreased, but measurable, enzyme activity may affect the ability of 5α-reductase to bind its testosterone substrate, or decrease affinity for the NADPH cofactor. The end result is insufficient conversion of testosterone to DHT. |
| **Diagnosis** | The diagnosis of 5aRD requires a careful physical examination and detailed family history. Laboratory findings include normal to high levels of testosterone, and low concentrations of DHT. In adults, a random blood sample may be adequate to determine the testosterone to DHT ratio; in younger chidren, blood samples should be obtained following human chorionic gonadotropin stimulation. Further characterization of the defect can be performed by assaying enzyme activity in cultured genital skin fibroblasts, and by investigations of the 5α-reductase gene. |
| **Counseling issues** | Although the appearance of affected newborns is predominantly female, the significant degree of virilization at puberty has led many 5aRD individuals to choose a male gender role on reaching adolescence. In one patient, fertility was achieved with the help of intracytoplasmic sperm injection and *in vitro* fertilization; this patient had a familial form of 5aRD, in which the defect in enzyme activity was mild. In general, sperm counts are low. |

**Notes**

Treatment with testosterone results in elevation of plasma DHT to the normal adult range, most likely due to $5\alpha$-reductase isoenzyme 1, which is expressed in humans only after birth. Despite the rise in DHT, penile growth response is often disappointing when testosterone treatment is initiated in adulthood. Treatment prior to the onset of puberty may result in a greater growth response.

**References**

Mendonca BB, Inacio M, Costa EM et al. Male pseudohermaphroditism due to steroid 5alpha-reductase 2 deficiency. Diagnosis, psychological evaluation, and management. *Medicine (Baltimore)* 1996;75:64–76.

Sultan C, Paris F, Terouanne B et al. Disorders linked to insufficient androgen action in male children. *Hum Reprod Update* 2001;7:314–22.

# Aromatase Deficiency

(also known as: female pseudohermaphroditism due to placental aromatase deficiency)

| | |
|---|---|
| **MIM** | 107910 |
| **Clinical features** | Deficient or absent placental and fetal hepatic aromatase activity prevents the conversion of fetal adrenal androgens to estrogen. The resulting androgen excess leads to masculinization of the female fetus, and, starting in the second trimester, virilization of the mother. At puberty, girls with aromatase deficiency have primary amenorrhea, lack of breast development, polycystic ovaries, and elevated gonadotropin levels. In men, growth and development are normal until late adolescence when failure of epiphyseal closure results in prolonged linear growth (past age 20 years) and very tall stature. Skeletal manifestations include delayed bone age and undermineralization. As in girls, gonadotropin levels are elevated. Aromatase deficiency has been documented in two men, both of whom had truncal obesity and insulin resistance. |
| **Primary tissue or gland affected** | Placenta. |
| **Other organs, tissues, or glands affected** | Brain, ovary and testis, adipose tissue, fetal liver, muscle, hair follicles, bone, pituitary gland, immune system. |
| **Age of onset** | Estrogen deficiency due to aromatase defects affects females *in utero*. In males, clinical manifestations are first noted in late adolescence or early adulthood. |
| **Epidemiology** | This is a rare disorder: only two males and six females have so far been described. Their geographic locations include North America, Europe, and Japan. |
| **Inheritance** | Autosomal recessive. |

| Chromosomal location | 15q21.1 |
|---|---|
| Genes | *CYP19* (aromatase, a member of the cytochrome p450 superfamily). |
| Mutational spectrum | Single base-pair substitutions and premature stop codons have been reported. |
| Effect of mutation | Aromatase catalyzes the last and irreversible step in the synthesis of estrogen compounds from androgens. The *CYP19* mutations found in patients with aromatase deficiency interfere with those regions of the protein that are critical for enzymatic activity. |
| Diagnosis | The diagnosis is supported by the clinical features described above, undetectable levels of estrone and estradiol, and elevated levels of gonadotropins. The common form of polycystic ovary syndrome (PCOS) shares some features with aromatase deficiency in women, including amenorrhea, variable degrees of virilization, obesity, insulin resistance, and polycystic ovaries. Clinical and laboratory evidence of estrogen deficiency in patients with aromatase deficiency should allow clinicians to easily distinguish women with PCOS from those with aromatase deficiency. The clinical presentation of aromatase deficiency in men is nearly identical to that of estrogen receptor-α deficiency. Men with aromatase deficiency show a rapid advancement of bone maturation and improvement in bone mineralization in response to low-dose estrogen treatment; men with estrogen receptor deficiency show no response, even to high-dose estrogen therapy. |
| Counseling issues | Estrogen treatment promotes breast development in girls, leads to the regression of ovarian cysts, and lowers gonadotropin levels. One of the two affected men had normal-sized testes; the other had small testes and was infertile. |

**Notes**

The presence of truncal obesity and insulin resistance in patients with this syndrome underscores the role of estrogens in regulating regional fat distribution. Aromatase deficiency has proven extremely useful in defining the physiologic importance of estrogens in men.

**References**

Grumbach MM, Auchus RJ. Estrogen: Consequences and implications of human mutations in synthesis and action. *J Clin Endocrinol Metab* 1999;84:4677–94 (Review).

Simpson ER. Genetic mutations resulting in loss of aromatase activity in humans and mice. *J Soc Gynecol Investig* 2000;7(1 Suppl.):S18–21 (Review).

# Aromatase Excess

(also known as: familial gynecomastia due to increased aromatase activity; increased activity of Cyp19)

| | |
|---|---|
| **MIM** | 107910 |

**Clinical features**  Prepubertal gynecomastia, rapid linear growth, and advanced skeletal maturation are the classical findings in boys with aromatase excess. Fertility in adult males appears to be normal; however, gonadotropin secretion may be mildly suppressed due to negative feedback by excess estrogens, resulting in decreased Leydig cell function and mild inhibition of testicular growth. Affected girls have isosexual precocious puberty, including advanced bone age. Combined central and peripheral precocious puberty may develop. Adult women with aromatase excess have exaggerated breast enlargement; sexual function, menstruation, and fertility appear to be normal.

**Primary tissue or gland affected**  Breast tissue.

**Other organs, tissues, or glands affected**  In addition to normal breast and breast cancer cells, the human p450 aromatase gene is expressed in the placenta, ovary, testis, brain, skin fibroblasts, adipocytes, and various fetal tissues.

**Age of onset**  During childhood, aromatase excess presents as gynecomastia in boys and isosexual precocious puberty in girls.

**Epidemiology**  This is a very rare condition. Few additional reports have appeared since the initial description in 1977.

**Inheritance**  The first reports of familial aromatase excess suggested X-linked inheritance. More recent studies and localization of the aromatase gene to Ch15q21 indicate an autosomal dominant pattern or genetic heterogeneity.

| | |
|---|---|
| **Chromosomal location** | 15q21.1 |
| **Genes** | *CYP19* (aromatase, a member of the cytochrome p450 superfamily). |
| **Mutational spectrum** | In contrast to aromatase deficiency, no specific mutations in *CYP19* have been identified in patients with aromatase excess. A sequence change in the p450 aromatase promoter region has been proposed as a likely candidate for the genetic defect in this syndrome. |
| **Effect of mutation** | A splicing error or altered binding of a transcription factor that regulates p450 aromatase mRNA synthesis seems likely. The net effect would increase the expression of p450 aromatase and result in increased enzymatic activity, leading to accelerated synthesis of estrone ($E_1$) and estradiol ($E_2$) from androstenedione and testosterone. |
| **Diagnosis** | Excess aromatase activity should be suspected when the clinical features of increased estrogen production are found, together with laboratory evidence of elevated $E_2$ and (especially) $E_1$ levels. A pubertal response to gonadotropin-releasing hormone (GnRH) stimulation does not exclude this diagnosis, since central precocious puberty may develop following a period of elevated peripheral estrogen synthesis. A positive family history is helpful in establishing the diagnosis of aromatase excess. Other diagnoses to consider include the common condition of adolescent gynecomastia, exogenous estrogen exposure, and estrogen-secreting tumors. Markedly elevated aromatase activity has been found in a feminizing adrenal adenoma, which caused isosexual precocious puberty in a young girl. |
| **Counseling issues** | The overall good prognosis for fertility should be emphasized. |
| **Notes** | When treatment is needed for combined central and peripheral precocious puberty, a GnRH analog to suppress gonadotropin synthesis, and an aromatase inhibitor to decrease peripheral |

estrogen synthesis, should be considered. Surgical treatment of gynecomastia or macromastia may be indicated.

**References**

Brodie A, Inkster S, Yue W. Aromatase expression in the human male. *Mol Cell Endocrinol* 2001;178:23–8.

Stratakis CA, Vottero A, Brodie A et al. The aromatase excess syndrome is associated with feminization of both sexes and autosomal dominant transmission of aberrant P450 aromatase gene transcription. *J Clin Endocrinol Metab* 1998;83:1348–57.

# Klinefelter's Syndrome

(also known as: KS; seminiferous tubule dysplasia)

| | |
|---|---|
| **MIM** | Not listed. |
| **Clinical features** | The most common features of KS are small testes and infertility in a phenotypic male. Most patients have gynecomastia, some show decreased facial and pubic hair, and, less often, penile size is small. Height is usually above average, and disproportionately long legs may be noticeable even before puberty. The mean IQ is in the 85–90 range, and severe retardation is uncommon. Intelligence is normal, but learning disabilities, including dyslexia, attention deficit disorders, and delays in speech and language acquisition are more common among KS patients. Mental retardation may occur in KS men with three or more X chromosomes in the karyotype. KS patients are at risk for decreased bone mineral density, development of type 2 diabetes mellitus, varicose veins, venous thromboses, and pulmonary embolism. Taurodontism (thinning of the tooth surface and hypertrophy of the pulp) occurs in nearly 40% of KS men and leads to early tooth decay. Compared with classical KS patients, phenotypic 46,XX males tend to be shorter, have fewer intellectual problems, and normal skeletal proportions. |
| **Primary tissue or gland affected** | Gonads. |
| **Age of onset** | Testicular biopsies have shown a reduction in the number of germ cells as early as infancy. Clinical manifestations usually appear in adolescence as delayed or incomplete pubertal development, small testes, and gynecomastia. In adult men, KS may be diagnosed during an evaluation of infertility. |
| **Epidemiology** | The prevalence of classical KS is approximately 1:500–1000 males. No specific geographic or ethnic variations have been observed. Advanced maternal age is associated with an increased incidence of |

KS. The 46,XX karyotype is found in approximately 1:20 000 phenotypic males.

**Inheritance**

Sporadic.

**Chromosomal location**

X

**Genes**

The 46,XX phenotypic male shares many clinical features with KS. Eighty percent of 46,XX phenotypic males are *SRY* positive. The gene responsible for classical KS is unknown.

**Mutational spectrum**

Classical KS is associated with a 47,XXY karyotype. Variant forms include: 46,XY/XXY mosaicism; 48,XXYY; 48,XXXY; 49,XXXYY; and 49,XXXXY. Features of KS have been observed in phenotypic males with a 46,XX karyotype resulting from a Y to X translocation.

**Effect of mutation**

Hyalinization and fibrosis of the seminiferous tubules and aggregation of Leydig cells results in azoospermia, variable degrees of testosterone deficiency, elevated gonadotropins, and gynecomastia.

**Diagnosis**

Routine amniocentesis may disclose a karyotype indicative of KS. School-age boys with small testes, learning problems, and other KS features, such as gynecomastia or eunuchoid body proportions, should be evaluated and referred for karyotyping. In prepubertal patients, LH and FSH levels may be normal. Although gonadotropins are typically elevated in KS, gonadotropin levels may be normal in patients with mosaic karyotypes. Testicular biopsy should be considered when mosaicism is suspected in a patient with a normal karyotype, and may be helpful in assessing the presence of viable germ cells.

**Counseling issues**

The incidence of breast carcinoma is approximately 20 times that found in normal men, and is associated with gynecomastia in nearly all cases. Extragonadal germ cell tumors are seen more frequently in KS patients, and typically appear in late adolescence or early adulthood with mediastinal metastatic disease. Men with KS show an increased incidence of systemic lupus, rheumatoid arthritis, and Sjogren syndrome. Considerable progress has been made in assisted reproduction using spermatozoa from the patient's own ejaculate or from testicular biopsies. Most of the offspring that result from assisted reproduction in nonmosaic KS men appear to be normal and have normal karyotypes.

**Notes**

Androgen replacement (usually parenteral testosterone esters, starting at low doses) should be administered when there is evidence of androgen deficiency, beginning in adolescence. Severe gynecomastia may require surgery.

**References**

Greco E, Rienzi L, Ubaldi F et al. Klinefelter's syndrome and assisted reproduction. *Fertil Steril* 2001;76:1068–9.

Smyth CM, Bremner WJ. Klinefelter syndrome. *Arch Intern Med* 1998;158:1309–14 (Review).

# Polycystic Ovary Syndrome

(also known as: PCOS; Stein–Leventhal syndrome; hyperandrogenism)

| | |
|---|---|
| MIM | 184700 |

**Clinical features**  PCOS is a complex clinical syndrome, the primary features of which include hirsutism, obesity, infertility, and enlarged, polycystic ovaries. However, none of these findings is diagnostic by itself, and women with PCOS may be lean, some may ovulate, and the ovarian morphology may not be found on standard investigation. As many as 20% of seemingly reproductively normal woman may have polycystic ovary morphology. PCOS patients have an apparent hypothalamic/pituitary defect, characterized by abnormal gonadotropin secretion, insulin resistance that cannot be explained on the basis of obesity alone, a β-cell defect that appears independent of insulin resistance, and increased ovarian androgen production. PCOS patients have a markedly increased risk of developing type 2 diabetes. Like other complex syndromes, such as type 2 diabetes itself, it is not clear which of these disorders is primary. Indeed, in different patients, different defects may be primary. Insulin resistance due to well-defined defects at the level of the insulin receptor can cause hyperandrogenism and PCOS ovarian morphology, and thus this may be the primary defect. Improved insulin sensitivity, using medication, diet, or weight loss, may improve hyperandrogenism and ovarian function. However, ovarian stromal cells that have been removed from patients with PCOS and cultured *in vitro* maintain their phenotype of increased androgen production, implying a primary ovarian defect. The male phenotype for PCOS is thought to be premature male-pattern baldness and/or excessive body hair.

**Primary tissue or gland affected**  Gonads.

**Age of onset**  Diagnosis is usually made in the second to fourth decades of life, with the onset of hyperandrogenism and menstrual irregularity.

| Epidemiology | The prevalence of PCOS among reproductive-age women has been estimated at 4%–12%, depending on the population studied and the diagnostic criteria used. About three quarters of women with anovulatory infertility and 90% of those with hirsuitism have PCOS. |
|---|---|
| Inheritance | Complex, autosomal dominant. |
| Chromosomal location | 15q23-q24 (CYP11A), 5q11.2 (FST). |
| Genes | *CYP11A* (cholesterol side-chain cleavage enzyme), *FST* (follistatin). |
| Mutational spectrum | These two genes have been identified by linkage analysis and association studies; however, specific mutations have not yet been described. PCOS appears to be an oligogenic syndrome, meaning that mutations in any of several genes can cause the syndrome. Studies suggest that, in some families, the disease is inherited as an autosomal dominant trait. In other families, a more complex mode of inheritance appears to be present, requiring the inheritance of mutations at more than one locus. Linkage and association studies have shown that genes at several other loci may be responsible for PCOS. In the UK population, the variable number of tandem repeats polymorphism near the insulin gene appears to be another good candidate for causing some cases of PCOS. |
| Effect of mutation | The syndrome is complex, and the primary defect is unknown. Some studies suggest that the primary defect is intrinsic to the ovary and results in increased ovarian androgen production; others propose that the primary defect results in resistance to insulin action, which causes secondary increased androgen production. Furthermore, some studies suggest a primary pituitary/hypothalamic defect. It is likely that any of these mechanisms, alone or in combination, may be active in any given individual. |
| Diagnosis | The formal diagnosis of PCOS is made according to a set of diagnostic criteria; however, different researchers use different |

criteria to define the syndrome. Most US investigators define PCOS as hyperandrogenism and chronic anovulation in the absence of specific diseases of the adrenal, ovary, or pituitary, such as hyperprolactinemia, 21-hydroxylase deficiency, or androgen-secreting tumors. This caveat is needed since the ovarian morphologic finding in PCOS (more than seven subcapsular follicular cysts <10 mm in diameter, and increased ovarian stroma) may occur as a secondary phenomenon in women with increased androgens from any source. Outside the US, diagnosis is usually based on ovarian morphology, with patients subgrouped according to ovulatory status.

**Counseling issues**

Although the syndrome may be inherited as an autosomal dominant trait, the clinical presentation appears to be highly dependent on environmental factors. Thus, a healthy diet, exercise, and maintenance of ideal body weight may prevent clinical disease in a genetically predisposed woman.

**Notes**

The genetics of PCOS is still unknown. As stated above, the findings to date appear to be most consistent with an oligogenic disorder. Mutations in different genes, perhaps affecting different metabolic pathways, can cause a similar clinical syndrome. Some mutations may be sufficient to cause the syndrome in the heterozygous state, and thus the syndrome is transmitted within a family as a dominant trait. Others may be insufficient to produce the clinical manifestations, and thus cause the syndrome only when inherited along with other minor genetic risk factors. Therefore, it is not surprising that different studies, using different populations, will frequently produce apparently conflicting results.

**References**

Dunaif A, Thomas A. Current concepts in the polycystic ovary syndrome. *Annu Rev Med* 2001;52:401–19 (Review).

Franks S, Gharani N, McCarthy M. Candidate genes in polycystic ovary syndrome. *Hum Reprod Update* 2001;7:405–10 (Review).

# 8. Abbreviations

| | |
|---|---|
| 17-OHP | 17 hydroxyprogesterone |
| 5aRD | 5α-reductase deficiency |
| AAA | achalasia–Addisonianism–alacrima |
| ABCD1 | ATP-binding cassette, subfamily D, member 1 |
| ACTH | adrenocorticotropic hormone |
| ADCC | adrenocortical carcinoma |
| ADHR | autosomal dominant hypophosphatemic rickets |
| ADP | adenosine diphosphate |
| AHC | adrenal hypoplasia, congenital |
| AHCH | adrenal hypoplasia with hypogonadotropic hypogonadism |
| AHO | Albright hereditary osteodystrophy |
| AIS | androgen-insensitivity syndrome |
| AITD | autoimmune thyroid disease |
| ALB | albumin |
| ALD | adrenoleukodystrophy |
| ALPL | alkaline phosphatase liver-type |
| AME | apparent mineralocorticoid excess |
| AMH | anti-Müllerian hormone |
| AMHR | anti-Müllerian hormone receptor |
| AMN | adrenomyeloneuropathy |
| AP | anteroposterior |
| APECED | autoimmune polyendocrinopathy-candidiasis-ectodermal dystrophy |
| APOE | apolipoprotein E |
| AQP | aquaporin |
| AR | androgen receptor |
| ATP | adenosine triphosphate |
| AVP | arginine vasopressin |
| BBS | Bardet–Biedl syndrome |
| BGP | bone Gla protein |
| BP | binding protein |
| BWS | Beckwith–Wiedemann syndrome |
| CAH | congenital adrenal hyperplasia |
| cAMP | cyclic adenosine monophosphate |
| CaSR | calcium-sensing receptor |

| | |
|---|---|
| cDI | central diabetes insipidus |
| CGD | chronic granulomatous disease |
| CMO | corticosterone methyl oxidase |
| CNC | Carney complex |
| CNS | central nervous system |
| CoA | coenzyme A |
| CPT | carnitine palmitoyltransferase |
| CT | computed tomography |
| DBD | DNA-binding domain |
| DDAVP | 1-desamino-8-D-arginine vasopressin |
| DDS | Denys–Drash syndrome |
| DFRX | drosophila fat facets related X |
| DHEA | dehydroepiandosterone |
| DHT | dihydrotestosterone |
| DI | diabetes insipidus |
| DIDMOAD | diabetes insipidus and diabetes mellitus with optic atrophy and deafness |
| DIT | di-iodotyrosine |
| DMD | Duchenne muscular dystrophy |
| DMSD | diabetes mellitus-secretory diarrhea syndrome |
| DMTN | diabetes mellitus, transient neonatal |
| DOC | 11-deoxycorticosterone |
| DPT | Diabetes Prevention Trial |
| $E_1$ | estrone |
| $E_2$ | estradiol |
| EGF | epidermal growth factor |
| EMG | exomphalos–macroglossia–gigantism syndrome |
| ENaC | epithelial sodium channel |
| ER | endoplasmic reticulum |
| FBH | familial benign hypercalcemia |
| FBHOk | familial benign hypercalcemia, Oklahoma variant |
| FBN | fibrillin |
| FDH | familial dysalbuminemic hyperthyroxinemia |
| FGD | familial glucocorticoid deficiency |
| FGF | fibroblast growth factor |
| FGFR | fibroblast growth factor receptor |

| | |
|---|---|
| FHH | familial hypocalciuric hypercalcemia |
| FIH | familial isolated hypoparathyroidism |
| FIHP | familial isolated primary hyperparathyroidism |
| FISH | fluorescence *in situ* hybridization |
| FMPP | familial male precocious puberty |
| FMTC | familial medullary thyroid carcinoma |
| FNMTC | familial nonmedullary thyroid carcinoma |
| FS | Frasier syndrome |
| FSH | follicle-stimulating hormone |
| G-6-Pase | glucose-6-phosphatase |
| GAD | glutamic acid decarboxylase |
| GCK | glucokinase |
| GH | growth hormone |
| GH-BP | growth hormone-binding protein |
| GHR | growth hormone receptor |
| GHRH | growth hormone-releasing hormone |
| GHRH-R | growth hormone-releasing hormone receptor |
| GK | glycerol kinase |
| GKD | glycerol kinase deficiency |
| GLUD | glutamate dehydrogenase |
| GnRH | gonadotropin-releasing hormone |
| GnRHR | gonadotropin-releasing hormone receptor |
| GRA | glucocorticoid-remediable aldosteronism |
| GRH | glucocorticoid-remediable hyperaldosteronism |
| GSH | glucocorticoid-suppressible hyperaldosteronism |
| GTP | guanosine 5′-triphosphate |
| hCG | human chorionic gonadotropin |
| HFE | hemochromatosis |
| HH | hereditary hemochromatosis |
| HHC | hypocalciuric hypercalcemia |
| HI | hyperinsulinism of infancy |
| HI/HA | hyperinsulinemia–hyperammonemia |
| HLA | human leukocyte antigen |
| HNF | hepatic nuclear factor |
| HRPT–JT | hyperparathyroidism–jaw tumor syndrome |
| HRPT | hyperparathyroidism |

| | |
|---|---|
| HSD | hydroxysteroid dehydrogenase |
| HYPX | hypoparathyroidism, X-linked |
| iACTHR | isolated ACTH-resistant |
| Ig | immunoglobulin |
| IGD | isolated gonadotropin deficiency |
| IGF | insulin-like growth factor |
| IGHD | isolated growth hormone deficiency |
| IHH | idiopathic hypogonadotropic hypogonadism |
| IL | interleukin |
| INS | insulin |
| INSR | insulin receptor |
| IPEX | immunodysregulation, polyendocrinopathy, enteropathy X-linked |
| IPF | insulin promoter factor |
| IRS | insulin receptor substrate |
| ISS | idiopathic short stature |
| IUGR | intrauterine growth retardation |
| JH | juvenile hemochromatosis |
| JT | jaw tumor |
| KATP | ATP-dependent potassium channel |
| KS | Klinefelter's syndrome |
| LADA | latent autoimmune diabetes in the adult |
| LBD | ligand-binding domain |
| LH | luteinizing hormone |
| LHR | luteinizing hormone receptor |
| LMD | Langer mesomelic dysplasia |
| LOH | loss of heterozygosity |
| LWD | Leri–Weill dyschondrosteosis |
| MADD | multiple acyl-coenzyme A dehydrogenase deficiency |
| MAS | McCune–Albright syndrome |
| MC2R | melanocortin 2 receptor |
| MCAD | medium-chain acyl-coenzyme A dehydrogenase |
| MCM | methylmalonyl CoA mutase |
| MEA | multiple endocrine adenomatosis |
| MEN | multiple endocrine neoplasia |
| MFS | Marfan's syndrome |

| | |
|---|---|
| MHC | major histocompatibility complex |
| MIBG | meta-iodobenzylguanidine |
| MIDD | maternally inherited diabetes and deafness |
| MIM | Mendelian Inheritance in Man |
| MIT | mono-iodotyrosine |
| MKKS | McKusick–Kaufman syndrome |
| MOD | maturity-onset diabetes |
| MODY | maturity-onset diabetes of the young |
| MPHD | multiple pituitary hormone deficiencies |
| MR | mineralocorticoid receptor |
| MRI | magnetic resonance imaging |
| MSUD | maple syrup urine disease |
| MTC | medullary thyroid carcinoma |
| mtDNA | mitochondrial DNA |
| NADPH | nicotinamide adenine dinucelotide phosphate |
| NAME | nevi, atrial myxoma, mucinosis of skin, endocrine hyperactivity |
| nDI | nephrogenic diabetes insipidus |
| NF | neurofibromatosis |
| NIDDM | noninsulin-dependent diabetes mellitus |
| NIS | sodium–iodide symporter |
| NMTC | nonmedullary thyroid cancer |
| NP | neurophysin |
| NS | Noonan's syndrome |
| NSHPT | neonatal severe primary hyperparathyroidism |
| OI | osteogenesis imperfecta |
| OTCD | ornithine transcarbamoylase deficiency |
| PAR | pseudoautosomal region |
| PCOS | polycystic ovary syndrome |
| PDS | Pendred syndrome |
| PGA | polyglandular autoimmune syndrome |
| PHA | pseudohypoaldosteronism |
| PHHI | persistent hyperinsulinemic hypoglycemia of infancy |
| PHP | pseudohypoparathyroidism |
| PHS | Pallister–Hall syndrome |
| PIT | pituitary-specific transcription factor |

| | |
|---|---|
| PKA | protein kinase A |
| PMDS | persistent Müllerian duct syndrome |
| POU1F1 | POU domain, class 1, transcription factor 1 |
| PPARG | peroxisome proliferator-activated receptor $\gamma$ |
| PPNAD | primary pigmented nodular adrenocortical disease |
| PROP1 | prophet of PIT1 |
| PTH | parathyroid hormone |
| PTHR | parathyroid hormone receptor |
| PTH-RP | parathyroid hormone-related protein |
| PWS | Prader–Willi syndrome |
| RCC | renal cell carcinoma |
| RP | retinitis pigmentosa |
| RSS | Russell–Silver syndrome |
| SADDAN | severe achondroplasia with developmental delay and acanthosis nigricans |
| SCAD | short-chain acyl-CoA dehydrogenase |
| SD | standard deviation |
| SH | succinate dehydrogenase |
| SE | standard error |
| serpin | serine protease inhibitor |
| SHOX | short-stature homeobox |
| SMMCI | solitary median maxillary central incisor |
| SOD | septo-optical dysplasia |
| StAR | steroidogenic acute regulatory protein |
| Stat | signal transducer and activator of transcription |
| SUR | sulfonylurea receptor |
| T:A | testosterone/androstenedione |
| T1DM | type 1 diabetes mellitus |
| T2DM | type 2 diabetes mellitus |
| T3 | tri-iodothyronine |
| T4 | thyroxine |
| TBG | thyroxine-binding globulin |
| TBG-CD | complete thyroxine-binding globulin deficiency |
| TD | thanatophoric dysplasia |
| TFR | transferrin receptor |
| TG | thyroglobulin |

| TGF | transforming growth factor |
| THOX | thyroid oxidase |
| THR | thyroid hormone receptor |
| TNDM | transient neonatal diabetes mellitus |
| TNSALP | tissue-nonspecific alkaline phosphatase |
| TPO | thyroid peroxidase |
| TRH | thyrotropin-releasing hormone |
| TRHR | thyrotropin-releasing hormone receptor |
| TRMA | thiamine-responsive megaloblastic anemia syndrome |
| TS | Turner's syndrome |
| TSC | thiazide-sensitive NaCl cotransporter |
| TSH | thyrotropin |
| TSHR | thyrotropin receptor |
| TTF | thyroid transcription factor |
| UPD | uniparental disomy |
| VDDR | vitamin D-dependent rickets |
| VDR | vitamin D receptor |
| VHL | von Hippel-Lindau |
| VLCFA | very long-chain fatty acid |
| VNTR | variable number tandem repeat |
| WAGR | Wilms' tumor, aniridia |
| WFS | wolframin |
| WS | Williams' syndrome |
| X-HypR | X-linked hypophosphatemic rickets |
| XLAAD | X-linked autoimmunity-allergic dysregulation syndrome |
| XPID | X-linked polyendocrinopathy, immune dysfunction and diarrhea |
| ZFX | zinc finger X |

# Glossary

# A

**Adenine (A)**

One of the bases making up **DNA** and **RNA** (pairs with **thymine** in DNA and **uracil** in RNA).

**Agarose gel electrophoresis**

See **electrophoresis**

**Allele**

One of two or more alternative forms of a **gene** at a given location (**locus**). A single allele for each locus is inherited separately from each parent. In normal human beings there are two alleles for each locus (**diploidy**). If the two alleles are identical, the individual is said to be **homozygous** for that allele; if different, the individual is **heterozygous**.

For example, the normal **DNA** sequence at **codon** 6 in the beta-globin gene is GAG (coding for glutamic acid), whereas in sickle cell disease the sequence is GTG (coding for valine). An individual is said to be heterozygous for the glutamic acid → valine **mutation** if he/she possesses one normal (GAG) and one mutated (GTG) allele. Such individuals are **carriers** of the sickle cell gene and do not manifest classical sickle cell disease (which is **autosomal recessive**).

**Allelic heterogeneity**

Similar/identical **phenotypes** caused by different **mutations** within a **gene**. For example, many different mutations in the same gene are now known to be associated with Marfan's syndrome (*FBN1* gene at 15q21.1).

**Amniocentesis**

Withdrawal of amniotic fluid, usually carried out during the second trimester, for the purpose of prenatal diagnosis.

**Amplification**

The production of increased numbers of a **DNA** sequence.

1. *In vitro*
In the early days of recombinant DNA techniques, the only way to amplify a sequence of interest (so that large amounts were available

for detailed study) was to **clone** the fragment in a vector (**plasmid** or phage) and transform bacteria with the recombinant vector. The transformation technique generally results in the 'acceptance' of a single vector molecule by each bacterial cell. The vector is able to exist autonomously within the bacterial cell, sometimes at very high copy numbers (e.g. 500 vector copies per cell). Growth of the bacteria containing the vector, coupled with a method to recover the vector sequence from the bacterial culture, allows for almost unlimited production of a sequence of interest. Cloning and bacterial propagation are still used for applications requiring either large quantities of material or else exceptionally pure material.

However, the advent of the **polymerase chain reaction** (PCR) has meant that amplification of desired DNA sequences can now be performed more rapidly than was the case with cloning (a few hours cf. days), and it is now routine to amplify DNA sequences 10 million-fold.

*2. In vivo*
Amplification may also refer to an increase in the number of DNA sequences within the genome. For example, the genomes of many tumors are now known to contain regions that have been amplified many fold compared to their nontumor counterparts (i.e. a sequence or region of DNA that normally occurs once at a particular chromosomal location may be present in hundreds of copies in some tumors). It is believed that many such regions harbor **oncogenes**, which, when present in high copy number, predispose to development of the malignant **phenotype**.

| | |
|---|---|
| **Aneuploid** | Possessing an incorrect number (abnormal complement) of **chromosomes**. The normal human complement is 46 chromosomes, any cell that deviates from this number is said to be aneuploid. |
| **Aneuploidy** | The chromosomal condition of a cell or organism with an incorrect number of **chromosomes**. Individuals with Down's syndrome are described as having aneuploidy, because they possess an extra copy of chromosome 21 (**trisomy** 21), making a total of 47 chromosomes. |

**Anticipation**

A general phenomenon that refers to the observation of an increase in severity, and/or decrease in age of onset, of a condition in successive generations of a family (see **Figure 1**). Anticipation is now known, in many cases, to result directly from the presence of a **dynamic mutation** in a family. In the absence of a dynamic mutation, anticipation may be explained by '**ascertainment bias**'. Thus, before the first dynamic mutations were described (in Fragile X and myotonic dystrophy), it was believed that ascertainment bias was the complete explanation for anticipation. There are two main reasons for ascertainment bias:

1. Identical **mutations** in different individuals often result in variable expressions of the associated **phenotype**. Thus, individuals within a family, all of whom harbor an identical mutation, may have variation in the severity of their condition.

2. Individuals with a severe phenotype are more likely to present to the medical profession. Moreover, such individuals are more likely to fail to reproduce (i.e. they are genetic lethals), often for social, rather than direct physical reasons.

For both reasons, it is much more likely that a mildly affected parent will be ascertained with a severely affected child, than the reverse. Therefore, the severity of a condition appears to increase through generations.

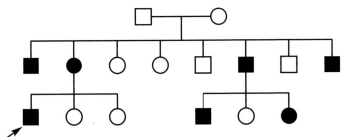

**Figure 1. Autosomal dominant inheritance** with **anticipation**. In many disorders that exhibit anticipation, the age of onset decreases in subsequent generations. It may happen that the transmitting parent (grandparent in this case) is unaffected at the time of presentation of the **proband** (see arrow). A good example is Huntington's disease, caused by the expansion of a CAG repeat in the coding region of the huntingtin gene. Note that this **pedigree** would also be consistent with either gonadal **mosaicism** or reduced **penetrance** (in the **carrier** grandparent).

*Genetics for Endocrinologists*

| | |
|---|---|
| **Anticodon** | The 3-base sequence on a **transfer RNA** (tRNA) molecule that is complementary to the 3-base **codon** of a **messenger RNA** (mRNA) molecule. |
| **Ascertainment bias** | See **anticipation** |
| **Autosomal disorder** | A disorder associated with a **mutation** in an autosomal **gene**. |
| **Autosomal dominant (AD) inheritance** | An **autosomal disorder** in which the **phenotype** is expressed in the **heterozygous** state. These disorders are not sex-specific. Fifty percent of offspring (when only one parent is affected) will usually manifest the disorder (see **Figure 2**). Marfan syndrome is a good example of an AD disorder; affected individuals possess one wild-type (normal) and one mutated **allele** at the *FBN1* **gene**. |

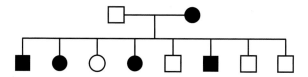

Figure 2. Autosomal dominant (AD) inheritance.

| | |
|---|---|
| **Autosomal recessive (AR) inheritance** | An **autosomal disorder** in which the **phenotype** is manifest in the **homozygous** state. This pattern of inheritance is not sex-specific and is difficult to trace through generations because both parents must contribute the abnormal **gene**, but may not necessarily display the disorder. The children of two **heterozygous** AR parents have a 25% chance of manifesting the disorder (see **Figure 3**). Cystic fibrosis (CF) is a good example of an AR disorder; affected individuals possess two **mutations**, one at each **allele**. |

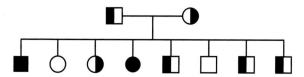

Figure 3. Autosomal recessive (AR) inheritance.

| Autosome | Any **chromosome**, other than the **sex chromosomes** (X or Y), that occurs in pairs in **diploid** cells. |
|---|---|

# B

| Barr body | An inactive **X chromosome**, visible in the **somatic cells** of individuals with more than one X chromosome (i.e. all normal females and all males with Klinefelter's syndrome). For individuals with nX chromosomes, n−1 Barr bodies are seen. The presence of a Barr body in cells obtained by **amniocentesis** or **chorionic villus sampling** used to be used as an indication of the sex of a baby before birth. |
|---|---|
| Base pair (bp) | Two **nucleotides** held together by hydrogen bonds. In **DNA, guanine** always pairs with **cytosine**, and **thymine** with **adenine**. A base pair is also the basic unit for measuring DNA length. |

# C

| Carrier | An individual who is **heterozygous** for a mutant **allele** (i.e. carries one wild-type (normal copy) and one mutated copy of the **gene** under consideration). |
|---|---|
| CentiMorgan (cM) | Unit of genetic distance. If the chance of **recombination** between two loci is 1%, the loci are said to be 1 cM apart. On average, 1 cM implies a physical distance of 1 Mb (1 000 000 **base pairs**) but significant deviations from this rule of thumb occur because recombination frequencies vary throughout the **genome**. Thus if recombination in a certain region is less likely than average, 1 cM may be equivalent to 5 Mb (5 000 000 base pairs) in that region. |
| Centromere | Central constriction of the **chromosome** where daughter **chromatids** are joined together, separating the short (p) from the long (q) arms (see **Figure 4**). |

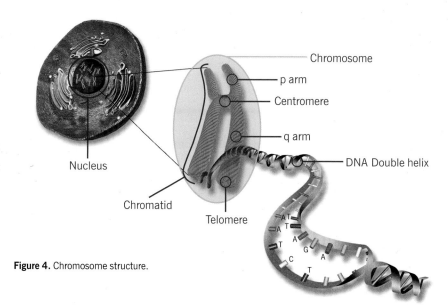

**Figure 4.** Chromosome structure.

| | |
|---|---|
| **Chorionic villus sampling (CVS)** | Prenatal diagnostic procedure for obtaining fetal tissue at an earlier stage of gestation than **amniocentesis**. Generally performed after 10 weeks, ultrasound is used to guide aspiration of tissue from the villus area of the chorion. |
| **Chromatid** | One of the two parallel identical strands of a **chromosome**, connected at the **centromere** during **mitosis** and **meiosis** (see **Figure 4**). Before replication, each chromosome consists of only one chromatid. After replication, two identical sister chromatids are present. At the end of mitosis or meiosis, the two sisters separate and move to opposite poles before the cell splits. |
| **Chromatin** | A readily stained substance in the nucleus of a cell consisting of **DNA** and proteins. During cell division it coils and folds to form the metaphase **chromosomes**. |
| **Chromosome** | One of the threadlike 'packages' of **genes** and other **DNA** in the nucleus of a cell (see **Figure 4**). Humans have 23 pairs of chromosomes, 46 in total: 44 **autosomes** and two **sex chromosomes**. Each parent contributes one chromosome to each pair. |

| **Chromosomal disorder** | A disorder that results from gross changes in **chromosome** dose. May result from addition or loss of entire chromosomes or just portions of chromosomes. |
|---|---|
| **Clone** | A group of genetically identical cells with a common ancestor. |
| **Codon** | A 3-base coding unit of **DNA** that specifies the function of a corresponding unit (**anticodon**) of **transfer RNA** (tRNA). |
| **Complementary DNA (cDNA)** | **DNA** synthesized from **messenger RNA** (mRNA) using **reverse transcriptase**. Differs from **genomic** DNA because it lacks **introns**. |
| **Complementation** | The wild-type **allele** of a **gene** compensates for a mutant allele of the same gene so that the heterozygote's **phenotype** is wild-type. |
| **Complementation analysis** | A genetic test (usually performed *in vitro*) that determines whether or not two **mutations** that produce the same **phenotype** are allelic. It enables the geneticist to determine how many distinct **genes** are involved when confronted with a number of mutations that have similar phenotypes. |
| | Occasionally it can be observed clinically. Two parents who both suffer from **recessive** deafness (i.e. both are **homozygous** for a mutation resulting in deafness) may have offspring that have normal hearing. If A and B refer to the wild-type (normal) forms of the genes, and a and b the mutated forms, one parent could be aa,BB and the other AA,bb. If **alleles** A and B are distinct, each child will have the **genotype** aA,bB and will have normal hearing. If A and B are allelic, the child will be homozygous at this **locus** and will also suffer from deafness. |
| **Compound heterozygote** | An individual with two different mutant **alleles** at the same **locus**. |
| **Concordant** | A pair of twins who manifest the same **phenotype** as each other. |

| Consanguinity | Sharing a common ancestor, and thus genetically related. **Recessive** disorders are seen with increased frequency in consanguineous families. |
|---|---|
| Consultand | An individual seeking genetic advice. |
| **Contiguous gene syndrome** | A syndrome resulting from the simultaneous functional imbalance of a group of **genes** (see **Figure 5**). The nomenclature for this group of disorders is somewhat confused, largely as a result of the history of their elucidation. The terms submicroscopic rearrangement/deletion/ duplication and microrearrangement/deletion/duplication are often used interchangeably. Micro or submicroscopic refer to the fact that such lesions are not detectable with standard cytogenetic approaches (where the limit of resolution is usually 10 Mb, and 5 Mb in only the most fortuitous of circumstances). A newer, and perhaps more comprehensive, term that is currently applied to this group of disorders is segmental aneusomy syndromes (SASs). This term embraces the possibility not only of loss or gain of a |

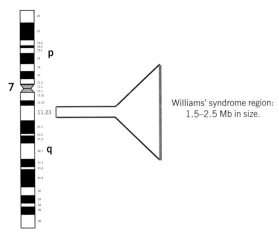

Williams' syndrome region: 1.5–2.5 Mb in size.

**Figure 5.** Schematic demonstrating the common deletion found in Williams' syndrome, at 7q11.23. The common deletion is not detectable using standard cytogenetic analysis (even high resolution), despite the fact that the deletion is at least 1.5 Mb in size. In practice, only genomic rearrangements that affect at least 5–10 Mb are detectable, either by standard cytogenetic analysis or, in fact, any technique whose endpoint involves analysis at the chromosomal level. Such deletions are termed microdeletions or submicroscopic deletions. Approximately 20 genes are known to be involved in the 7q11.23 microdeletion, and work is underway to determine which genes contribute to which aspects of the Williams' syndrome phenotype.

chromosomal region that harbors many genes (leading to imbalance of all those genes), but also of functional imbalance in a group of genes, as a result of an abnormality of the machinery involved in their silencing/**transcription** (i.e. methylation-based mechanisms that depend on a master control gene).

In practice, most contiguous gene syndromes result from the **heterozygous** deletion of a segment of **DNA** that is large in molecular terms but not detectable cytogenetically. The size of such deletions is usually 1.5–3.0 Mb. It is common for one to two dozen genes to be involved in such deletions, and the resultant **phenotypes** are often complex, involving multiple organ systems and, almost invariably, learning difficulties. A good example of a contiguous gene syndrome is Williams' syndrome, a sporadic disorder that is due to a heterozygous deletion at **chromosome** 7q11.23. Affected individuals have characteristic phenotypes, including recognizable facial appearance and typical behavioral traits (including moderate learning difficulties). Velocardiofacial syndrome is currently the most common **microdeletion** known, and is caused by deletions of 3 Mb at chromosome 22q11.

| | |
|---|---|
| **Crossing over** | Reciprocal exchange of genetic material between **homologous chromosomes** at **meiosis** (see **Figure 6**). |
| **Cytogenetics** | The study of the structure of **chromosomes**. |
| **Cytosine (C)** | One of the bases making up **DNA** and **RNA** (pairs with **guanine**). |
| **Cytotrophoblast** | Cells obtained from fetal chorionic villi by **chorionic villus sampling (CVS)**. Used for **DNA** and **chromosome** analysis. |

# D

| | |
|---|---|
| **Deletion** | A particular kind of **mutation** that involves the loss of a segment of **DNA** from a **chromosome** with subsequent re-joining of the two extant ends. It can refer to the removal of one or more bases within a **gene** or to a much larger aberration involving millions of bases. The |

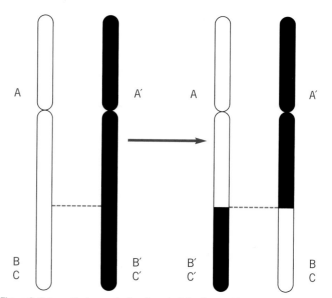

**Figure 6.** Schematic demonstrating the principle of **recombination (crossing over)**. On average, 50 recombinations occur per meiotic division (1–2 per **chromosome**). **Loci** that are far apart on the chromosome are more likely to be separated during recombination than those that are physically close to each other (they are said to be linked, see **linkage**), i.e. A and B are less likely to cosegregate than B and C. Note that the two **homologues** of a sequence have been differentially labeled according to their chromosome of origin.

term deletion is not totally specific, and differentiation must be made between **heterozygous** and **homozygous** deletions. Large heterozygous deletions are a common cause of complex **phenotypes** (see **contiguous gene syndrome**); large germ-line homozygous deletions are extremely rare, but have been described. Homozygous deletions are frequently described in **somatic cells**, in association with the manifestation of the malignant phenotype. The two deletions in a homozygous deletion need not be identical, but must result in the complete absence of DNA sequences that occupy the 'overlap' region.

**Denature**

Broadly used to describe two general phenomena:

1. The 'melting' or separation of double-stranded **DNA** (dsDNA) into its constituent single strands, which may be achieved using heat or chemical approaches.

2. The denaturation of proteins. The specificity of proteins is a result of their 3-dimensional conformation, which is a function of their (linear) amino acid sequence. Heat and/or chemical approaches may result in denaturation of a protein—the protein loses its 3-dimensional conformation (usually irreversibly) and, with it, its specific activity.

**Diploid**

The number of **chromosomes** in most human **somatic cells** (46). This is double the number found in **gametes** (23, the **haploid** number).

**Discordant**

A pair of twins who differ in their manifestation of a **phenotype**.

**Dizygotic**

The fertilization of two separate eggs by two separate sperm resulting in a pair of genetically nonidentical twins.

**DNA**
**(deoxyribonucleic acid)**

The molecule of heredity. DNA normally exists as a double-stranded (ds) molecule; one strand is the complement (in sequence) of the other. The two strands are joined together by hydrogen bonding, a noncovalent mechanism that is easily reversible using heat or chemical means. DNA consists of four distinct bases: **guanine** (G), **cytosine** (C), **thymine** (T), and **adenine** (A). The convention is that DNA sequences are written in a 5′ to 3′ direction, where 5′ and 3′ refer to the numbering of carbons on the deoxyribose ring. A guanine on one strand will always pair with a cytosine on the other strand, while thymine pairs with adenine. Thus, given the sequence of bases on one strand, the sequence on the other is immediately determined:

5′–AGTGTGACTGATCTTGGTG–3′
3′–TCACACTGACTAGAACCAC–5′

The complexity (informational content) of a DNA molecule resides almost completely in the particular sequence of its bases. For a sequence of length 'n' **base pairs**, there are 4n possible sequences. Even for relatively small n, this number is astronomical ($4n = 1.6 \times 10^{60}$ for n = 100).

The complementarity of the two strands of a dsDNA molecule is a very important feature and one that is exploited in almost all molecular genetic techniques. If dsDNA is **denatured**, either by heat or by chemical means, the two strands become separated from each other. If the conditions are subsequently altered (e.g. by reducing heat), the two strands eventually 'find' each other in solution and re-anneal to form dsDNA once again. The specificity of this reaction is quite high, under the right circumstances—strands that are not highly complementary are much less likely to re-anneal compared to perfect or near perfect matches. The process by which the two strands 'find' each other depends on random molecular collisions, and a '**zippering**' mechanism, which is initiated from a short stretch of complementarity. This property of DNA is vital for the **polymerase chain reaction (PCR)**, **Southern blotting**, and any method that relies on the use of a DNA/**RNA probe** to detect its counterpart in a complex mix of molecules.

**DNA chip**

A 'chip' or microarray of multiple **DNA** sequences immobilized on a solid surface (see **Figure 7**). The term chip refers more often to semiconductor-based DNA arrays, in which short DNA sequences (oligos) are synthesized *in situ*, using a photolithographic process akin to that used in the manufacture of semiconductor devices for the electronics industry. The term microarray is much more general and includes any collection of DNA sequences immobilized onto a solid surface, whether by a photolithographic process, or by simple 'spotting' of DNA sequences onto glass slides.

The power of DNA microarrays is based on the parallel analysis that they allow for. In conventional **hybridization** analysis (i.e. **Southern blotting**), a single DNA sequence is usually used to interrogate a small number of different individuals. In DNA microarray analysis, this approach is reversed—an individual's DNA is hybridized to an array that may contain 30 000 distinct spots. This allows for direct information to be obtained about all DNA sequences on the array in one experiment. DNA microarrays have been used successfully to directly uncover **point mutations** in single **genes**, as well as detect

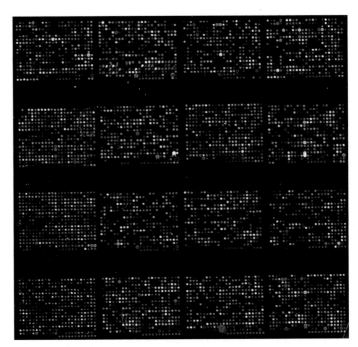

**Figure 7.** DNA chip. DNA arrays (or 'chips') are composed of thousands of 'spots' of DNA, attached to a solid surface (normally glass). Each spot contains a different DNA sequence. The arrays allow for massively parallel experiments to be performed on samples. In practice, two samples are applied to the array. One sample is a control (from a 'normal' sample) and one is the test sample. Each sample is labeled with fluorescent tags, control with green and test with red. The two labeled samples are cohybridized to the array and the results read by a laser scanner. Spots on the array whose DNA content is equally represented in the test and control samples yield equal intensities in the red and green channels, resulting in a yellow signal. Spots appearing as red represent DNA sequences that are present at higher concentration in the test sample compared to the control sample and vice versa.

alterations in **gene expression** associated with certain disease states/cellular differentiation. It is likely that certain types of array will be useful in the determination of subtle copy number alterations, as occurs in **microdeletion/microduplication** syndromes.

**DNA methylation**  Addition of a methyl group ($-CH_3$) to **DNA nucleotides** (often **cytosine**). Methylation is often associated with reduced levels of expression of a given **gene** and is important in **imprinting**.

| DNA replication | Use of existing **DNA** as a template for the synthesis of new DNA strands. In humans and other eukaryotes, replication takes place in the cell nucleus. DNA replication is semiconservative—each new double-stranded molecule is composed of a newly synthesized strand and a pre-existing strand. |
|---|---|

| Dominant (traits/diseases) | Manifesting a **phenotype** in the **heterozygous** state. Individuals with Huntington's disease, a dominant condition, are affected even though they possess one normal copy of the **gene**. |
|---|---|

| Dynamic/ nonstable mutation | The vast majority of **mutations** known to be associated with human genetic disease are intergenerationally stable (no alteration in the mutation is observed when transmitted from parent to child). However, a recently described and growing class of disorders result from the presence of mutations that are unstable intergenerationally. These disorders result from the presence of tandem repeats of short **DNA** sequences (e.g. the sequence CAG may be repeated many times in tandem), see **Table 1**. For reasons that are not completely clear, the copy number of such repeats may vary from parent to child (usually resulting in a copy number increase) and within the **somatic cells** of a given individual. Abnormal **phenotypes** result when the number of repeats reaches a given threshold. Furthermore, when this threshold has been reached, the risk of even greater expansion of copy number in subsequent generations increases. |
|---|---|

# E

| Electrophoresis | The separation of molecules according to size and ionic charge by an electrical current.

*Agarose gel electrophoresis*
Separation, based on size, of **DNA/RNA** molecules through agarose. Conventional agarose gel electrophoresis generally refers to electrophoresis carried out under standard conditions, allowing the resolution of molecules that vary in size from a few hundred to a few thousand **base pairs**. |
|---|---|

| Disorder | Protein/location | Repeat | Repeat location | Normal range | Pre-mutation | Full mutation | Type | MIM |
|---|---|---|---|---|---|---|---|---|
| Progressive myoclonus epilepsy of Unverricht-Lundborg type (EPM1) | cystatin B 21q22.3 | $C_4GC_4G$ CG | Promoter | 2–3 | 12–17 | 30–75 | AR | 254800 |
| Fragile X type A (FRAXA) | FMR1 Xq27.3 | CGG | 5'UTR | 6–52 | ~60–200 | ~200–>2,000 | XLR | 309550 |
| Fragile X type E (FRAXE) | FMR2 Xq28 | CGG 5 | C'UTR | 6–25 | – | >200 | XLR | 309548 |
| Friedreich's ataxia (FRDA) | frataxin 9q13 | GAA | intron | 17–22 | – | 200–>900 | AR | 229300 |
| Huntington's disease (HD) | huntingtin 4p16.3 | CAG | ORF | 6–34 | – | 36–180 | AD | 143100 |
| Dentatorubal-pallidoluysian atrophy (DRPLA) | atrophin 12p12 | CAG | ORF | 7–25 | – | 49–88 | AD | 125370 |
| Spinal and bulbar muscular atrophy (SBMA – Kennedy syndrome) | androgen receptor Xq11-12 | CAG | ORF | 11–24 | – | 40–62 | XLR | 313200 |
| Spinocerebellar ataxia type 1 (SCA1) | ataxin-1 6p23 | CAG | ORF | 6–39 | – | 39–83 | AD | 164400 |
| Spinocerebellar ataxia type 2 (SCA2) | ataxin-2 12q24 | CAG | ORF | 15–29 | – | 34–59 | AD | 183090 |
| Spinocerebellar ataxia type 3 (SCA3) | ataxin-3 14q24.3-q31 | CAG | ORF | 13–36 | – | 55–84 | AD | 109150 |
| Spinocerebellar ataxia type 6 (SCA6) | PQ calcium channel 19p13 | CAG | ORF | 4–16 | – | 21–30 | AD | 183086 |
| Spinocerebellar ataxia type 7 (SCA7) | ataxin-7 3p21.1-p12 | CAG | ORF | 4–35 | 28–35 | 34–>300 | AD | 164500 |
| Spinocerebellar ataxia type 8 (SCA8) | SCA8 13q21 | CTG | 3'UTR | 6–37 | – | ~107–250[1] | AD | 603680 |
| Spinocerebellar ataxia type 10 (SCA10) | SCA10 22q13-qter | ATTCT | intron 9 | 10–22 | – | 500–4,500 | AD | 603516 |
| Spinocerebellar ataxia type 12 (SCA12) | PP2R2B 5q31-33 | CAG | 5'UTR | 7–28 | – | 66–78 | AD | 604326 |
| Myotonic dystrophy (DM) | DMPK 19q13.3 | CTG | 3'UTR | 5–37 | ~50–180 | ~200–>2,000 | AD | 160900 |

**Table 1.** 'Classical' repeat expansion disorders. [1]Longer alleles exist but are not associated with disease. AD: autosomal dominant; AR: autosomal recessive; ORF: open reading frame (coding region); 3′ UTR: 3′ untranslated region (downstream of gene); 5′ UTR: 5′ untranslated region (upstream of gene); XLR: X-linked recessive.

*Polyacrylamide gel electrophoresis*
Allows resolution of proteins or DNA molecules differing in size by only 1 base pair.

*Pulsed field gel electrophoresis*
(Also performed using agarose) refers to a specialist technique that allows resolution of much larger DNA molecules, in some cases up to a few Mb in size.

**Empirical recurrence risk – recurrence risk**

Based on observation, rather than detailed knowledge of, e.g., modes of inheritance or environmental factors.

**Endonuclease**

An enzyme that cleaves **DNA** at an internal site (see also **restriction enzyme**).

**Euchromatin**

**Chromatin** that stains lightly with trypsin G banding and contains active/potentially active **genes**.

**Euploidy**

Having a normal **chromosome** complement.

**Exon**

Coding part of a **gene**. Historically, it was believed that all of a **DNA** sequence is mirrored exactly on the messenger **RNA** (mRNA) molecule (except for the presence of **uracil** in mRNA compared to **thymine** in DNA). It was a surprise to discover that this is generally not the case. The **genomic** sequence of a gene has two components: **exons** and **introns**. The exons are found in both the genomic sequence and the mRNA, whereas the introns are found only in the genomic sequence. The mRNA for dystrophin, an **X-linked** gene associated with Duchenne muscular dystrophy (DMD), is 14 000 **base pairs** long but the genomic sequence is spread over a distance of 1.5 million base pairs, because of the presence of very long intronic sequences. After the genomic sequence is initially transcribed to RNA, a complex system ensures specific removal of introns. This system is known as **splicing**.

**Expressivity**

Degree of expression of a disease. In some disorders, individuals carrying the same **mutation** may manifest wide variability in severity of the disorder. **Autosomal dominant** disorders are often associated with **variable expressivity**, a good example being Marfan's syndrome. Variable expressivity is to be differentiated from **incomplete penetrance**, an all or none phenomenon that refers to the complete absence of a **phenotype** in some **obligate carriers**.

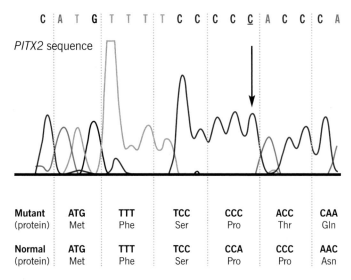

| C | A | T | G | T | T | T | T | C | C | C | C | C | A | C | C | A |

*PITX2* sequence

| | | | | | |
|---|---|---|---|---|---|
| **Mutant** (protein) | ATG Met | TTT Phe | TCC Ser | CCC Pro | ACC Thr | CAA Gln |
| **Normal** (protein) | ATG Met | TTT Phe | TCC Ser | CCA Pro | CCC Pro | AAC Asn |

**Figure 8. Frame-shift mutation**. This example shows a sequence of *PITX2* in a patient with Rieger's syndrome, an **autosomal dominant** condition. The sequence graph shows only the abnormal sequence. The arrow indicates the insertion of a single **cytosine** (C) residue. When translated, the triplet code is now out of frame by 1 base pair. This totally alters the translated protein's amino acid sequence. This leads to a premature **stop codon** later in the protein and results in Rieger's syndrome.

# F

**Familial**  Any trait that has a higher frequency in relatives of an affected individual than the general population.

**FISH**  Fluorescence *in situ* hybridization (see **In situ hybridization**).

**Founder effect**  The high frequency of a mutant **allele** in a population as a result of its presence in a founder (ancestor). Founder effects are particularly noticeable in relative genetic isolates, such as the Finnish or Amish.

**Frame-shift mutation**  **Deletion/insertion** of a **DNA** sequence that is not an exact multiple of 3 **base pairs**. The result is an alteration of the reading frame of the **gene** such that all sequence that lies beyond the **mutation** is missense (i.e. codes for the wrong amino acids) (see **Figure 8**). A premature **stop codon** is usually encountered shortly after the frame shift.

# G

**Gamete (germ cell)**  The mature male or female reproductive cells, which contain a **haploid** set of **chromosomes**.

**Gene**  An ordered, specific sequence of **nucleotides** that controls the transmission and expression of one or more traits by specifying the sequence and structure of a particular protein or **RNA** molecule. Mendel defined a gene as the basic physical and functional unit of all heredity.

**Gene expression**  The process of converting a **gene's** coded information into the existing, operating structures in the cell.

**Gene mapping**  Determines the relative positions of **genes** on a **DNA** molecule and plots the genetic distance in **linkage** units (**centiMorgans**) or physical distance (**base pairs**) between them.

**Genetic code**  Relationship between the sequence of bases in a nucleic acid and the order of amino acids in the polypeptide synthesized from it (see **Table 2**). A sequence of three nucleic acid bases (a triplet) acts as a codeword (**codon**) for one amino acid or instruction (start/stop).

**Genetic counseling**  Information/advice given to families with, or at risk of, genetic disease. Genetic counseling is a complex discipline that requires accurate diagnostic approaches, up-to-date knowledge of the genetics of the condition, an insight into the beliefs/anxieties/wishes of the individual seeking advice, intelligent risk estimation, and, above all, skill in communicating relevant information to individuals from a wide variety of educational backgrounds. Genetic counseling is most often carried out by trained medical geneticists or, in some countries, specialist genetic counselors or nurses.

**Genetic heterogeneity**  Association of a specific **phenotype** with **mutations** at different loci. The broader the phenotypic criteria, the greater the heterogeneity

| | | 2nd | 2nd | 2nd | 2nd | | |
|---|---|---|---|---|---|---|---|
| | | T | C | A | G | | |
| 1st | T | TTT Phe [F] | TCT Ser [S] | TAT Tyr [Y] | TGT Cys [C] | T | 3rd |
| | | TTC Phe [F] | TCC Ser [S] | TAC Tyr [Y] | TGC Cys [C] | C | |
| | | TTA Leu [L] | TCA Ser [S] | TAA Ter [end] | TGA Ter [end] | A | |
| | | TTG Leu [L] | TCG Ser [S] | **TAG Ter [end]** | TGG Trp [W] | G | |
| 1st | C | CTT Leu [L] | CCT Pro [P] | CAT His [H] | CGT Arg [R] | T | 3rd |
| | | CTC Leu [L] | CCC Pro [P] | CAC His [H] | CGC Arg [R] | C | |
| | | CTA Leu [L] | CCA Pro [P] | CAA Gln [Q] | CGA Arg [R] | A | |
| | | CTG Leu [L] | CCG Pro [P] | CAG Gln [Q] | CGG Arg [R] | G | |
| 1st | A | ATT Ile [I] | ACT Thr [T] | AAT Asn [N] | AGT Ser [S] | T | 3rd |
| | | ATC Ile [I] | ACC Thr [T] | AAC Asn [N] | AGC Ser [S] | C | |
| | | ATA Ile [I] | ACA Thr [T] | AAA Lys [K] | AGA Arg [R] | A | |
| | | ATG Met [M] | ACG Thr [T] | AAG Lys [K] | AGG Arg [R] | G | |
| 1st | G | GTT Val [V] | GCT Ala [A] | GAT Asp [D] | GGT Gly [G] | T | 3rd |
| | | GTC Val [V] | GCC Ala [A] | GAC Asp [D] | GGC Gly [G] | C | |
| | | GTA Val [V] | GCA Ala [A] | GAA Glu [E] | GGA Gly [G] | A | |
| | | GTG Val [V] | GCG Ala [A] | GAG Glu [E] | GGG Gly [G] | G | |

**Table 2.** The **genetic code**. To locate a particular codon (e.g. TAG, marked in bold) locate the first base (T) in the left hand column, then the second base (A) by looking at the top row, and finally the third (G) in the right hand column (TAG is a stop codon). Note the redundancy of the genetic code—for example, three different codons specify a stop signal, and threonine (Thr) is specified by any of ACT, ACC, ACA, and ACG.

(e.g. mental retardation). However, even very specific phenotypes may be genetically heterogeneous. Isolated central hypothyroidism is a good example: this **autosomal recessive** condition is now known to be associated (in different individuals) with mutations in the TSH β chain at 1p13, the TRH receptor at 8q23, or TRH itself at 3q13.3–q21. There is no obvious distinction between the clinical phenotypes associated with these two genes. Genetic heterogeneity should not be confused with **allelic heterogeneity**, which refers to the presence of different mutations at the same **locus**.

**Genetic locus**

A specific location on a **chromosome**.

**Genetic map**

A map of genetic landmarks deduced from **linkage (recombination) analysis**. Aims to determine the linear order of a set of **genetic markers** along a **chromosome**. Genetic maps differ significantly from **physical maps**, in that recombination frequencies are not identical across different **genomic** regions, resulting occasionally in large discrepancies.

| | |
|---|---|
| **Genetic marker** | A **gene** that has an easily identifiable **phenotype** so that one can distinguish between those cells or individuals that do or do not have the gene. Such a gene can also be used as a **probe** to mark cell nuclei or **chromosomes**, so that they can be isolated easily or identified from other nuclei or chromosomes later. |
| **Genetic screening** | Population analysis designed to ascertain individuals at risk of either suffering or transmitting a genetic disease. |
| **Genetically lethal** | Prevents reproduction of the individual, either because the condition causes death prior to reproductive age, or because social factors make it highly unlikely (although not impossible) that the individual concerned will reproduce. |
| **Genome** | The complete **DNA** sequence of an individual, including the **sex chromosomes** and **mitochondrial DNA** (mtDNA). The genome of humans is estimated to have a complexity of $3.3 \times 10^9$ **base pairs** (per **haploid** genome). |
| **Genomic** | Pertaining to the **genome**. Genomic **DNA** differs from **complementary DNA** (cDNA) in that it contains noncoding as well as coding DNA. |
| **Genotype** | Genetic constitution of an individual, distinct from expressed features (**phenotype**). |
| **Germ line** | Germ cells (those cells that produce **haploid gametes**) and the cells from which they arise. The germ line is formed very early in embryonic development. Germ line **mutations** are those present constitutionally in an individual (i.e. in all cells of the body) as opposed to somatic mutations, which affect only a proportion of cells. |
| **Giemsa banding** | Light/dark bar code obtained by staining **chromosomes** with Giemsa stain. Results in a unique bar code for each chromosome. |
| **Guanine (G)** | One of the bases making up **DNA** and **RNA** (pairs with **cytosine**). |

# H

**Haploid**

The **chromosome** number of a normal **gamete**, containing one each of every individual chromosome (23 in humans).

**Haploinsufficiency**

The presence of one active copy of a **gene**/region is insufficient to compensate for the absence of the other copy. Most genes are not 'haploinsufficient'—50% reduction of gene activity does not lead to an abnormal **phenotype**. However, for some genes, most often those involved in early development, reduction to 50% often correlates with an abnormal phenotype. Haploinsufficiency is an important component of most **contiguous gene disorders** (e.g. in Williams' syndrome, **heterozygous deletion** of a number of genes results in the mutant phenotype, despite the presence of normal copies of all affected genes).

**Hemizygous**

Having only one copy of a **gene** or **DNA** sequence in **diploid** cells. Males are hemizygous for most genes on the **sex chromosomes**, as they possess only one **X chromosome** and one **Y chromosome** (the exceptions being those genes with counterparts on both sex chromosomes). **Deletions** on **autosomes** produce hemizygosity in both males and females.

**Heterochromatin**

Contains few active **genes**, but is rich in highly repeated simple sequence **DNA**, sometimes known as satellite DNA. Heterochromatin refers to inactive regions of the **genome**, as opposed to **euchromatin**, which refers to active, gene expressing regions. Heterochromatin stains darkly with **Giemsa**.

**Heterozygous**

Presence of two different **alleles** at a given **locus**.

**Histones**

Simple proteins bound to **DNA** in **chromosomes**. They help to maintain **chromatin** structure and play an important role in regulating **gene** expression.

| | |
|---|---|
| **Holandric** | Pattern of inheritance displayed by **mutations** in **genes** located only on the **Y chromosome**. Such mutations are transmitted only from father to son. |
| **Homologue or homologous gene** | Two or more **genes** whose sequences manifest significant similarity because of a close evolutionary relationship. May be between species (orthologues) or within a species (paralogues). |
| **Homologous chromosomes** | **Chromosomes** that pair during **meiosis**. These chromosomes contain the same linear **gene** sequences as one another and derive from one parent. |
| **Homology** | Similarity in **DNA** or protein sequences between individuals of the same species or among different species. |
| **Homozygous** | Presence of identical **alleles** at a given **locus**. |
| **Human gene therapy** | The study of approaches to treatment of human genetic disease, using the methods of modern molecular genetics. Many trials are under way studying a variety of disorders, including cystic fibrosis. Some disorders are likely to be more treatable than others—it is probably going to be easier to replace defective or absent **gene** sequences rather than deal with genes whose aberrant expression results in an actively toxic effect. |
| **Human genome project** | Worldwide collaboration aimed at obtaining a complete sequence of the human **genome**. Most sequencing has been carried out in the USA, although the Sanger Centre in Cambridge, UK has sequenced one third of the genome, and centers in Japan and Europe have also contributed significantly. The first draft of the human genome was released in the summer of 2000 to much acclaim. The finished sequence may not be available until 2003. Celera, a privately funded venture, headed by Dr Craig Ventner, also published its first draft at the same time. |

| | |
|---|---|
| **Hybridization** | Pairing of complementary strands of nucleic acid. Also known as **re-annealing**. May refer to re-annealing of **DNA** in solution, on a membrane (**Southern blotting**) or on a DNA microarray. May also be used to refer to fusion of two **somatic cells**, resulting in a hybrid that contains genetic information from both donors. |

# I

| | |
|---|---|
| **Imprinting** | A general term used to describe the phenomenon whereby a **DNA** sequence (coding or otherwise) carries a signal or imprint that indicates its parent of origin. For most DNA sequences, no distinction can be made between those arising paternally and those arising maternally (apart from subtle sequence variations); for imprinted sequences this is not the case. The mechanistic basis of imprinting is almost always **methylation**—for certain **genes**, the copy that has been inherited from the father is methylated, while the maternal copy is not. The situation may be reversed for other imprinted genes. Note that imprinting of a gene refers to the general phenomenon, not which parental copy is methylated (and, therefore, usually inactive). Thus, formally speaking, it is incorrect to say that a gene undergoes paternal imprinting. It is correct to say that the gene undergoes imprinting and that the inactive (methylated) copy is always the paternal one. However, in common genetics parlance, paternal imprinting is usually understood to mean the same thing. |
| ***In situ* hybridization (ISH)** | Annealing of **DNA** sequences to immobilized **chromosomes**/cells/tissues. Historically done using radioactively labeled **probes**, this is currently most often performed with fluorescently tagged molecules (fluorescent *in situ* hybridization – **FISH**, see **Figure 9**). ISH/FISH allows for the rapid detection of a DNA sequence within the **genome**. |
| **Incomplete penetrance** | Complete absence of expression of the abnormal **phenotype** in a proportion of individuals known to be **obligate carriers**. To be distinguished from **variable expressivity**, in which the phenotype always manifests in obligate carriers, but with widely varying degrees of severity. |

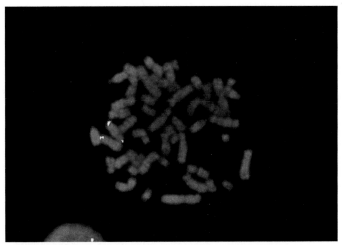

**Figure 9. Fluorescence *in situ* hybridization. FISH** analysis of a patient with a complex syndrome, using a clone containing **DNA** from the region 8q24.3. In addition to that clone, a control from 8pter was used. The 8pter clone has yielded a signal on both **homologues** of **chromosome** 8, while the 'test' clone from 8q24.3 has yielded a signal on only one homologue, demonstrating a (**heterozygous**) deletion in that region.

**Index case – proband**    The individual through which a family medically comes to light. For example, the index case may be a baby with Down's syndrome. Can be termed propositus (if male) or proposita (if female).

**Insertion**    Interruption of a chromosomal sequence as a result of insertion of material from elsewhere in the **genome** (either a different **chromosome**, or elsewhere from the same chromosome). Such insertions may result in abnormal **phenotypes** either because of direct interruption of a **gene** (uncommon), or because of the resulting imbalance (i.e. increased dosage) when the chromosomes that contain the normal counterparts of the inserted sequence are also present.

**Intron**    A noncoding **DNA** sequence that 'interrupts' the protein-coding sequences of a **gene**; intron sequences are transcribed into **messenger RNA** (mRNA), but are cut out before the mRNA is translated into a protein (this process is known as **splicing**). Introns may contain sequences involved in regulating expression of a gene. Unlike the

**exon**, the intron is the **nucleotide** sequence in a gene that is not represented in the amino acid sequence of the final gene product.

**Inversion**

A structural abnormality of a **chromosome** in which a segment is reversed, as compared to the normal orientation of the segment. An inversion may result in the reversal of a segment that lies entirely on one chromosome arm (paracentric) or one that spans (i.e. contains) the **centromere** (pericentric). While individuals who possess an inversion are likely to be genetically balanced (and therefore usually phenotypically normal), they are at increased risk of producing unbalanced offspring because of problems at **meiosis** with pairing of the inversion chromosome with its normal **homologue**. Both **deletions** and duplications may result, with concomitant congenital abnormalities related to **genomic** imbalance, or miscarriage if the imbalance is lethal.

# K

**Karyotype**

A photomicrograph of an individual's **chromosomes** arranged in a standard format showing the number, size, and shape of each chromosome type, and any abnormalities of chromosome number or morphology (see **Figure 10**).

**Kilobase (kb)**

1000 **base pairs** of **DNA**.

**Knudson hypothesis**

See **tumor suppressor gene**

# L

**Linkage**

Coinheritance of **DNA** sequences/**phenotypes** as a result of physical proximity on a **chromosome**. Before the advent of molecular genetics, linkage was often studied with regard to proteins, enzymes, or cellular characteristics. An early study demonstrated linkage between the Duffy blood group and a form of **autosomal dominant** congenital cataract (both are now known to reside at 1q21.1).

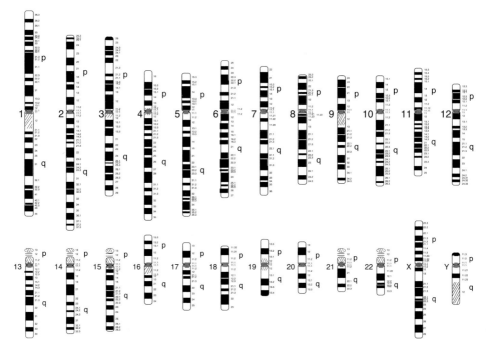

**Figure 10.** Schematic of a normal human (male) **karyotype**. (ISCN 550 ideogram produced by the MRC Human Genetics Unit, Edinburgh, reproduced with permission.)

Phenotypes may also be linked in this manner (i.e. families manifesting two distinct Mendelian disorders).

During the **recombination** phase of **meiosis**, genetic material is exchanged (equally) between two **homologous chromosomes**. **Genes/DNA sequences** that are located physically close to each other are unlikely to be separated during recombination. Sequences that lie far apart on the same chromosome are more likely to be separated. For sequences that reside on different chromosomes, segregation will always be random, so that there will be a 50% chance of two markers being coinherited.

| | |
|---|---|
| **Linkage analysis** | An algorithm designed to map (i.e. physically locate) an unknown **gene** (associated with the **phenotype** of interest) to a chromosomal |

region. Linkage analysis has been the mainstay of disease-associated gene identification for some years. The general availability of large numbers of DNA markers that are variable in the population (**polymorphisms**), and which therefore permit **allele** discrimination, has made linkage analysis a relatively rapid and dependable approach (see **Figure 11**). However, the method relies on the ascertainment of large families manifesting Mendelian disorders. Relatively little phenotypic heterogeneity is tolerated, as a single misassigned individual (believed to be unaffected despite being a gene **carrier**) in a **pedigree** may completely invalidate the results. **Genetic heterogeneity** is another problem, not within families (usually) but between families. Thus, conditions that result in identical phenotypes despite being associated with **mutations** within different genes (e.g. tuberous sclerosis) are often hard to study. Linkage analysis typically follows a standard algorithm:

1. Large families with a given disorder are ascertained. Detailed clinical evaluation results in assignment of affected vs. unaffected individuals.

2. Large numbers of polymorphic DNA markers that span the **genome** are analyzed in all individuals (affected and unaffected).

3. The results are analyzed statistically, in the hope that one of the markers used will have demonstrably been coinherited with the phenotype in question more often than would be predicted by chance.

The LOD score (**logarithm of the odds**) gives an indication of the likelihood of the result being significant (and not having occurred simply as a result of chance coinheritance of the given marker with the condition).

**Linkage disequilibrium**   Association of particular **DNA** sequences with each other, more often than is likely by chance alone (see **Figure 12**). Of particular relevance to inbred populations (e.g. Finland), where specific disease **mutations** are found to reside in close proximity to specific variants of DNA markers, as a result of the **founder effect**.

*Genetics for Endocrinologists*

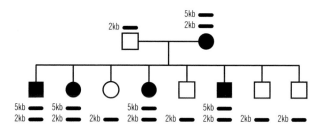

In the example above, note that the (affected) mother has a 5-kb band in addition to a 2-kb band. All the unaffected individuals have the small band only, all those who are affected have the large band. The unaffected individuals must have the mother's 2-kb fragment rather than her 5-kb fragment, and the affected individuals must have inherited the 5-kb band from the mother (as the father does not have one)—note that those individuals who only show the 2-kb band still have two alleles (one from each parent), they are just the same size and so cannot be differentiated. Thus, it appears that the 5-kb band is segregating with the disorder. The results in a family such as this are suggestive but further similar results in other families would be required for a sufficiently high **LOD** score.

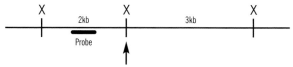

The **probe** recognizes a **DNA** sequence adjacent to a restriction site (see arrow) that is polymorphic (present on some **chromosomes** but not others). When such a site is present, the **DNA** is cleaved at that point and the probe detects a 2-kb fragment. When absent, the DNA is not cleaved and the probe detects a fragment of size (2 + 3) kb = 5 kb. X refers to the points at which the **restriction enzyme** will cleave the DNA. The recognition sequence for most restriction enzymes is very stringent—change in just one **nucleotide** will result in failure of cleavage. Most RFLPs result from the presence of a single nucleotide polymorphism that has altered the restriction site.

**Figure 11.** Schematic demonstrating the use of **restriction fragment length polymorphisms** (RFLPs) in **linkage analysis**.

**Linkage map**        A map of **genetic markers** as determined by genetic analysis (i.e. **recombination** analysis). May differ markedly from a map determined by actual physical relationships of genetic markers, because of the variability of recombination.

**Locus**        The position of a **gene/DNA** sequence on the **genetic map**. Allelic genes/sequences are situated at identical loci in **homologous chromosomes**.

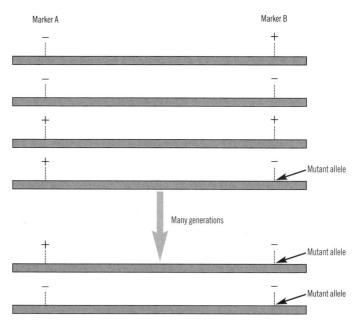

**Figure 12.** Schematic demonstrating the concept of **linkage disequilibrium**.

A **gene** is physically very close to marker B and further from marker A. Markers A and B, both on the same **chromosome**, can exist in one of two forms : +/−. Thus there are four possible **haplotypes**, as shown. If the **founder** mutation in the gene occurred as shown, then it is likely that even after many generations the mutant **allele** will segregate with the − form of marker B, as **recombination** is unlikely to have occurred between the two. However, since marker A is further away, the gene will now often segregate with the − form of marker A, which was not present on the original chromosome. The likelihood of recombination between the gene and marker A will depend on the physical distance between them, and on rates of recombination. It is possible that the gene would show a lesser but still significant degree of linkage disequilibrium with marker A.

**Locus heterogeneity**

**Mutations** at different loci cause similar **phenotypes**.

**LOD (Logarithm of the Odds) score**

A statistical test of **linkage**. Used to determine whether a result is likely to have occurred by chance or to truly reflect linkage. The LOD score is the logarithm (base 10) of the likelihood that the linkage is meaningful. A LOD score of 3 implies that there is only a 1:1000 chance that the results have occurred by chance (i.e. the result would be likely to occur once by chance in 1000 simultaneous

*Genetics for Endocrinologists*

studies addressing the same question). This is taken as proof of linkage (see **Figure 11**).

**Lyonization**

The inactivation of n–1 **X chromosomes** on a random basis in an individual with n X chromosomes. Named after Mary Lyon, this mechanism ensures dosage compensation of **genes** encoded by the X chromosome. X chromosome inactivation does not occur in normal males who possess only one X chromosome, but does occur in one of the two X chromosomes of normal females. In males who possess more than one X chromosome (i.e. XXY, XXXY, etc.), the rule is the same and only one X chromosome remains active. X-inactivation occurs in early embryonic development and is random in each cell. The inactivation pattern in each cell is faithfully maintained in all daughter cells. Therefore, females are genetic **mosaics**, in that they possess two populations of cells with respect to the X chromosome: one population has one X active, while in the other population the other X is active. This is relevant to the expression of **X-linked** disease in females.

# M

**Meiosis**

The process of cell division by which male and female **gametes** (germ cells) are produced. Meiosis has two main roles. The first is **recombination** (during meiosis I). The second is reduction division. Human beings have 46 **chromosomes**, and each is conceived as a result of the union of two germ cells; therefore, it is reasonable to suppose that each germ cell will contain only 23 chromosomes (i.e. the **haploid** number). If not, then the first generation would have 92 chromosomes, the second 184, etc. Thus, at meiosis I, the number of chromosomes is reduced from 46 to 23.

**Mendelian inheritance**

Refers to a particular pattern of inheritance, obeying simple rules: each **somatic cell** contains two **genes** for every characteristic and each pair of genes divides independently of all other pairs at **meiosis**.

| Mendelian Inheritance in Man (MIM/OMIM) | A catalogue of human Mendelian disorders, initiated in book form by Dr Victor McKusick of Johns Hopkins Hospital in Baltimore, USA. The original catalogue (produced in the mid-1960s) listed approximately 1500 conditions. By December 1998, this number had risen to 10 000; at the time of writing (February 2003) the figure had reached 14 158. With the advent of the Internet, MIM is now available as an online resource, free of charge (OMIM – Online Mendelian Inheritance in Man). The URL for this site is: http://www.ncbi.nlm.nih.gov/omim/. The online version is updated regularly, far faster than is possible for the print version; therefore, new **gene** discoveries are quickly assimilated into the database. OMIM lists disorders according to their mode of inheritance: |
|---|---|

1 ---- (100000- ) **Autosomal dominant** (entries created before May 15, 1994)

2 ---- (200000- ) **Autosomal recessive** (entries created before May 15, 1994)

3 ---- (300000- ) **X-linked** loci or **phenotypes**

4 ---- (400000- ) Y-linked loci or phenotypes

5 ---- (500000- ) Mitochondrial loci or phenotypes

6 ---- (600000- ) Autosomal loci/phenotypes (entries created after May 15, 1994).

Full explanations of the best way to search the catalogue are available at the home page for OMIM.

| Messenger RNA (mRNA) | The template for protein synthesis, carries genetic information from the nucleus to the ribosomes where the code is translated into protein. Genetic information flows: **DNA** $\rightarrow$ **RNA** $\rightarrow$ protein. |
|---|---|
| Methylation | See **DNA methylation** |
| Microdeletion | Structural **chromosome** abnormality involving the loss of a segment that is not detectable using conventional (even high resolution) |

cytogenetic analysis. Microdeletions usually involve 1–3 Mb of sequence (the resolution of cytogenetic analysis rarely is better than 10 Mb). Most microdeletions are **heterozygous**, although some individuals/families have been described with **homozygous** microdeletions. See also **contiguous gene syndrome**.

**Microduplication**  Structural **chromosome** abnormality involving the gain of a segment that may involve long sequences (commonly 1–3 Mb), which are, nevertheless, undetectable using conventional cytogenetic analysis. Patients with microduplications have three copies of all sequences within the duplicated segment, as compared to two copies in normal individuals. See also **contiguous gene syndrome**.

**Microsatellites**  **DNA** sequences composed of short tandem repeats (STRs), such as di- and trinucleotide repeats, distributed widely throughout the **genome** with varying numbers of copies of the repeating units. Microsatellites are very valuable as **genetic markers** for mapping human **genes**.

**Missense mutation**  Single base substitution resulting in a **codon** that specifies a different amino acid than the wild-type.

**Mitochondrial disease/disorder**  Ambiguous term referring to disorders resulting from abnormalities of mitochondrial function. Two separate possibilities should be considered.

1. **Mutations** in the mitochondrial **genome** (see **Figure 13**). Such disorders will manifest an inheritance pattern that mirrors the manner in which mitochondria are inherited. Therefore, a mother will transmit a mitochondrial mutation to all her offspring (all of whom will be affected, albeit to a variable degree). A father will not transmit the disorder to any of his offspring.

2. Mutations in nuclear encoded **genes** that adversely affect mitochondrial function. The mitochondrial genome does not code for all the genes required for its maintenance; many are encoded in the

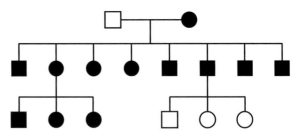

**Figure 13. Mitochondrial inheritance.** This **pedigree** relates to mutations in the mitochondrial **genome**.

nuclear genome. However, the inheritance patterns will differ markedly from the category described in the first option, and will be indistinguishable from standard Mendelian disorders.

Each mitochondrion possesses between 2–10 copies of its genome, and there are approximately 100 mitochondria in each cell. Therefore, each cell possesses 200–1000 copies of the mitochondrial genome. Heteroplasmy refers to the variability in sequence of this large number of genomes—even individuals with mitochondrial genome mutations are likely to have wild-type **alleles**. Variability in the proportion of molecules that are wild-type may have some bearing on the clinical variability often seen in such disorders.

| | |
|---|---|
| **Mitochondrial DNA** | The **DNA** in the circular **chromosome** of mitochondria. Mitochondrial DNA is present in multiple copies per cell and mutates more rapidly than **genomic** (nuclear) DNA. |
| **Mitosis** | Cell division occurring in **somatic cells**, resulting in two daughter cells that are genetically identical to the parent cell. |
| **Monogenic trait** | Causally associated with a single **gene**. |
| **Monosomy** | Absence of one of a pair of **chromosomes**. |
| **Monozygotic** | Arising from a single **zygote** or fertilized egg. Monozygotic twins are genetically identical. |

| | |
|---|---|
| **Mosaicism or mosaic** | Refers to the presence of two or more distinct cell lines, all derived from the same **zygote**. Such cell lines differ from each other as a result of **DNA** content/sequence. Mosaicism arises when the genetic alteration occurs postfertilization (postzygotic). The important features that need to be considered in mosaicism are: |

The proportion of cells that are 'abnormal'. In general, the greater the proportion of cells that are abnormal, the greater the severity of the associated **phenotype**.

The specific tissues that contain high levels of the abnormal cell line(s). This variable will clearly also be relevant to the manifestation of any phenotype. An individual may have a **mutation** bearing cell line in a tissue where the mutation is largely irrelevant to the normal functioning of that tissue, with a concomitant reduction in phenotypic sequelae.

Mosaicism may be functional, as in normal females who are mosaic for activity of the two **X chromosomes** (see **Lyonization**).

Mosaicism may occasionally be observed directly. **X-linked** skin disorders, such as incontinentia pigmenti, often manifest **mosaic** changes in the skin of a female, such that abnormal skin is observed alternately with normal skin, often in streaks (Blaschko's lines), which delineate developmental histories of cells.

| | |
|---|---|
| **Multifactorial inheritance** | A type of hereditary pattern resulting from a complex interplay of genetic and environmental factors. |
| **Mutation** | Any heritable change in **DNA** sequence. |

# N

| | |
|---|---|
| **Nondisjunction** | Failure of two **homologous chromosomes** to pull apart during **meiosis** I, or two **chromatids** of a chromosome to separate in meiosis II or **mitosis**. The result is that both are transmitted to one daughter cell, while the other daughter cell receives neither. |

| Nondynamic (stable) mutations | Stably inherited **mutations**, in contradistinction to **dynamic mutations**, which display variability from generation to generation. Includes all types of stable mutation (single base substitution, small **deletions/ insertions**, **microduplications**, and **microdeletions**). |
| --- | --- |
| Nonpenetrance | Failure of expression of a **phenotype** in the presence of the relevant **genotype**. |
| Nonsense mutation | A single base substitution resulting in the creation of a **stop codon** (see **Figure 14**). |
| Northern blot | **Hybridization** of a radiolabeled **RNA/DNA probe** to an immobilized RNA sequence. So called in order to differentiate it from **Southern blotting**, which was described first. Neither has any relationship to points on the compass. Southern blotting was named after its inventor Ed Southern |
| Nucleotide | A basic unit of **DNA** or **RNA** consisting of a nitrogenous base— **adenine**, **guanine**, **thymine**, or **cytosine** in DNA, and adenine, guanine, **uracil**, or cytosine in RNA. A nucleotide is composed of a phosphate molecule and a sugar molecule—deoxyribose in DNA and ribose in RNA. Many thousands or millions of nucleotides link to form a DNA or RNA molecule. |

# O

| Obligate carrier | See **obligate heterozygote** |
| --- | --- |
| Obligate heterozygote (obligate carrier) | An individual who, on the basis of **pedigree** analysis, must carry the mutant **allele**. |
| Oncogene | A **gene** that, when over expressed, causes neoplasia. This contrasts with **tumor suppressor genes**, which result in tumorigenesis when their activity is reduced. |

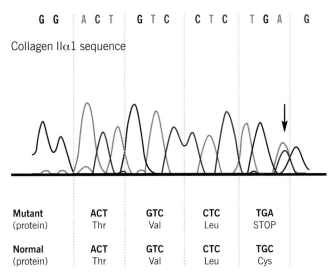

**Figure 14.** Nonsense mutation. This example shows a sequence graph of collagen II (α1) in a patient with Stickler syndrome, an **autosomal dominant** condition. The sequence is of **genomic DNA** and shows both normal and abnormal sequences (the patient is heterozygous for the mutation).
The base marked with an arrow has been changed from C to A. When translated the codon is changed from TGC (cysteine) to TGA (stop). The premature **stop codon** in the collagen gene results in Stickler syndrome.

# P

**p**

Short arm of a **chromosome** (from the French *petit*) (see **Figure 4**).

**Palindromic sequence**

A **DNA** sequence that contains the same 5′ to 3′ sequence on both strands. Most **restriction enzymes** recognize palindromic sequences. An example is 5′–AGATCT–3′, which would read 3′–TCTAGA–5′ on the complementary strand. This is the recognition site of *Bgl*II.

**Pedigree**

A schematic for a family indicating relationships to the **proband** and how a particular disease or trait has been inherited (see **Figure 15**).

**Penetrance**

An all-or-none phenomenon related to the proportion of individuals with the relevant **genotype** for a disease who actually manifest

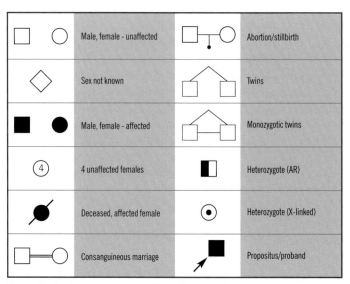

**Figure 15.** Symbols commonly used in **pedigree** drawing.

the **phenotype**. Note the difference between penetrance and **variable expressivity**.

**Phenotype**

Observed disease/abnormality/trait. An all-embracing term that does not necessarily imply pathology. A particular phenotype may be the result of **genotype**, the environment or both.

**Physical map**

A map of the locations of identifiable landmarks on **DNA**, such as specific DNA sequences or **genes**, where distance is measured in **base pairs**. For any **genome**, the highest resolution map is the complete **nucleotide** sequence of the **chromosomes**. A physical map should be distinguished from a **genetic map**, which depends on **recombination** frequencies.

**Plasmid**

Found largely in bacterial and protozoan cells, plasmids are autonomously replicating, extrachromosomal, circular **DNA** molecules that are distinct from the normal bacterial **genome** and are often used as vectors in recombinant DNA technologies. They

| | |
|---|---|
| | are not essential for cell survival under nonselective conditions, but can be incorporated into the genome and are transferred between cells if they encode a protein that would enhance survival under selective conditions (e.g. an enzyme that breaks down a specific antibiotic). |
| **Pleiotropy** | Diverse effects of a single **gene** on many organ systems (e.g. the **mutation** in Marfan's syndrome results in lens dislocation, aortic root dilatation, and other pathologies). |
| **Ploidy** | The number of sets of **chromosomes** in a cell. Human cells may be **haploid** (23 chromosomes, as in mature sperm or ova), **diploid** (46 chromosomes, seen in normal **somatic cells**), or triploid (69 chromosomes, seen in abnormal somatic cells, which results in severe congenital abnormalities). |
| **Point mutation** | Single base substitution. |
| **Polygenic disease** | Disease (or trait) that results from the simultaneous interaction of multiple **gene** mutations, each of which contributes to the eventual **phenotype**. Generally, each **mutation** in isolation is likely to have a relatively minor effect on the phenotype. Such disorders are not inherited in a Mendelian fashion. Examples include hypertension, obesity, and diabetes. |
| **Polymerase chain reaction (PCR)** | A molecular technique for amplifying **DNA** sequences *in vitro* (see **Figure 16**). The DNA to be copied is **denatured** to its single strand form and two synthetic oligonucleotide primers are annealed to complementary regions of the target DNA in the presence of excess deoxynucleotides and a heat-stable DNA polymerase. The power of PCR lies in the exponential nature of **amplification**, which results from repeated cycling of the 'copying' process. Thus, a single molecule will be copied in the first cycle, resulting in two molecules. In the second cycle, each of these will also be copied, resulting in four copies. In theory, after n cycles, there will be $2^n$ molecules for |

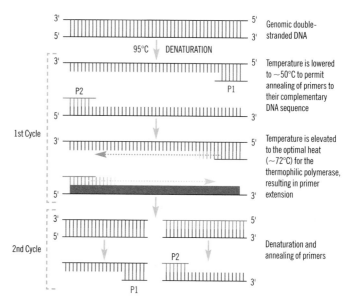

**Figure 16.** Schematic illustrating the technique of **polymerase chain reaction** (PCR).

each starting molecule. In practice, this theoretical limit is rarely reached, mainly for technical reasons. PCR has become a standard technique in molecular biology research as well as routine diagnostics.

**Polymorphism**

May be applied to **phenotype** or **genotype**. The presence in a population of two or more distinct variants, such that the frequency of the rarest is at least 1% (more than can be explained by recurrent **mutation** alone). A **genetic locus** is polymorphic if its sequence exists in at least two forms in the population.

**Premutation**

Any **DNA mutation** that has little, if any, phenotypic consequence but predisposes future generations to the development of full mutations with phenotypic sequelae. Particularly relevant in the analysis of diseases associated with **dynamic mutations**.

**Proband (propositus) – index case**

The first individual to present with a disorder through which a **pedigree** can be ascertained.

**Probe**

General term for a molecule used to make a measurement. In molecular genetics, a probe is a piece of **DNA** or **RNA** that is labeled and used to detect its complementary sequence (e.g. **Southern blotting**).

**Promoter region**

The noncoding sequence upstream (5′) of a **gene** where **RNA** polymerase binds. **Gene expression** is controlled by the promoter region both in terms of level and tissue specificity.

**Protease**

An enzyme that digests other proteins by cleaving them into small fragments. Proteases may have broad specificity or only cleave a particular site on a protein or set of proteins.

**Protease inhibitor**

A chemical that can inhibit the activity of a **protease**. Most proteases have a corresponding specific protease inhibitor.

**Proto-oncogene**

A misleading term that refers to **genes** that are usually involved in signaling and cell development, and are often expressed in actively dividing cells. Certain **mutations** in such genes may result in malignant transformation, with the mutated genes being described as **oncogenes**. The term proto-oncogene is misleading because it implies that such genes were selected for by evolution in order that, upon mutation, cancers would result because of oncogenic activation. A similar problem arises with the term **tumor suppressor gene**.

**Pseudogene**

Near copies of true **genes**. Pseudogenes share sequence **homology** with true genes, but are inactive as a result of multiple **mutations** over a long period of time.

**Purine**

A nitrogen-containing, double-ring, basic compound occurring in nucleic acids. The purines in **DNA** and **RNA** are **adenine** and **guanine**.

| Pyrimidine | A nitrogen-containing, single-ring, basic compound that occurs in nucleic acids. The pyrimidines in **DNA** are **cytosine** and **thymine**, and cytosine and **uracil** in **RNA**. |

# Q

| q | Long arm of a **chromosome** (see **Figure 4**). |

# R

| Re-annealing | See **hybridization** |

| Recessive (traits, diseases) | Manifest only in homozygotes. For the **X chromosome**, recessivity applies to males who carry only one (mutant) **allele**. Females who carry **X-linked mutations** are generally heterozygotes and, barring unfortunate X-inactivation, do not manifest X-linked recessive **phenotypes**. |

| Reciprocal translocation | The exchange of material between two non-**homologous chromosomes**. |

| Recombination | The creation of new combinations of linked **genes** as a result of **crossing over** at **meiosis** (see **Figure 6**). |

| Recurrence risk | The chance that a genetic disease, already present in a member of a family, will recur in that family and affect another individual. |

| Restriction enzyme | **Endonuclease** that cleaves double-stranded (ds)**DNA** at specific sequences. For example, the enzyme *Bg*III recognizes the sequence AGATCT, and cleaves after the first A on both strands. Most restriction endonucleases recognize sequences that are palindromic— the complementary sequence to AGATCT, read in the same orientation, is also AGATCT. The term 'restriction' refers to the function of these enzymes in nature. The organism that synthesizes a given restriction |

enzyme (e.g. *Bg*III) does so in order to 'kill' foreign DNA—
'restricting' the potential of foreign DNA that has become integrated
to adversely affect the cell. The organism protects its own DNA from
the restriction enzyme by simultaneously synthesizing a specific
methylase that recognizes the same sequence and modifies one of
the bases, such that the restriction enzyme is no longer able to
cleave. Thus, for every restriction enzyme, it is likely that a
corresponding methylase exists, although in practice only a relatively
small number of these have been isolated.

| | |
|---|---|
| **Restriction fragment length polymorphism (RFLP)** | A restriction fragment is the length of **DNA** generated when DNA is cleaved by a **restriction enzyme**. Restriction fragment length varies when a **mutation** occurs within a restriction enzyme sequence. Most commonly the **polymorphism** is a single base substitution, but it may also be a variation in length of a DNA sequence due to **variable number tandem repeats** (VNTRs). The analysis of the fragment lengths after DNA is cut by restriction enzymes is a valuable tool for establishing **familial** relationships and is often used in forensic analysis of blood, hair, or semen (see **Figure 11**). |
| **Restriction map** | A **DNA** sequence map, indicating the position of restriction sites. |
| **Reverse genetics** | Identification of the causative **gene** for a disorder, based purely on molecular genetic techniques, when no knowledge of the function of the gene exists (the case for most genetic disorders). |
| **Reverse transcriptase** | Catalyses the synthesis of **DNA** from a single-stranded **RNA** template. Contradicted the central dogma of genetics (DNA → RNA → protein) and earned its discoverers the Nobel Prize in 1975. |
| **RNA (ribonucleic acid)** | RNA molecules differ from **DNA** molecules in that they contain a ribose sugar instead of deoxyribose. There are a variety of types of RNA (including **messenger RNA, transfer RNA**, and ribosomal RNA) and they work together to transfer information from DNA to the protein-forming units of the cell. |

| Robertsonian translocation | A **translocation** between two acrocentric **chromosomes**, resulting from centric fusion. The short arms and satellites (chromosome segments separated from the main body of the chromosome by a constriction and containing highly repetitive **DNA**) are lost. |

# S

| Second hit hypothesis | See **tumor suppressor gene** |

| Sex chromosomes | Refers to the **X** and **Y chromosomes**. All normal individuals possess 46 chromosomes, of which 44 are **autosomes** and two are sex chromosomes. An individual's sex is determined by his/her complement of sex chromosomes. Essentially, the presence of a Y chromosome results in the male **phenotype**. Males have an X and a Y chromosome, while females possess two X chromosomes. The Y chromosome is small and contains relatively few **genes**, concerned almost exclusively with sex determination and/or sperm formation. By contrast, the X chromosome is a large chromosome that possesses many hundreds of genes. |

| Sex-limited trait | A trait/disorder that is almost exclusively limited to one sex and often results from **mutations** in autosomal **genes**. A good example of a sex-limited trait is breast cancer. While males are affected by breast cancer, it is much less common (~1%) than in women. Females are more prone to breast cancer than males, not only because they possess significantly more breast tissue, but also because their hormonal milieu is significantly different. In many cases, early onset bilateral breast cancer is associated with mutations either in *BRCA1* or *BRCA2*, both autosomal genes. An example of a sex-limited trait in males is male pattern baldness, which is extremely rare in premenopausal women. The inheritance of male pattern baldness is consistent with **autosomal dominant**, not **sex-linked dominant**, inheritance. |

| Sex-linked dominant | See **X-linked dominant** |

| Sex-linked recessive | See **X-linked recessive** |
|---|---|
| Sibship | All the sibs in a family. |
| Silent mutation | One that has no (apparent) phenotypic effect. |
| Single gene disorder | A disorder resulting from a **mutation** on one **gene**. |
| Somatic cell | Any cell of a multicellular organism not involved in the production of **gametes**. |
| Southern blot | **Hybridization** with a radiolabeled **RNA/DNA probe** to an immobilized DNA sequence (see **Figure 17**). Named after Ed Southern (currently Professor of Biochemistry at Oxford University, UK), the technique has spawned the nomenclature for other types of blot (**Northern blots** for RNA and **Western blots** for proteins). |

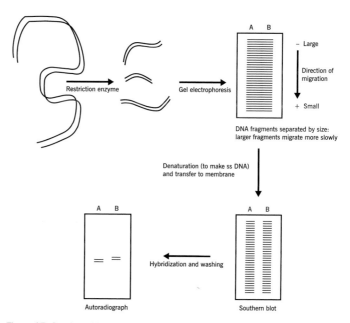

**Figure 17. Southern blotting.**

| Splicing | Removal of **introns** from precursor **RNA** to produce **messenger RNA** (mRNA). The process involves recognition of **intron–exon** junctions and specific removal of intronic sequences, coupled with reconnection of the two strands of **DNA** that formerly flanked the intron. |
| --- | --- |
| Start codon | The AUG **codon** of **messenger RNA** recognized by the ribosome to begin protein production. |
| Stop codon | The **codons** UAA, UGA, or UAG on **messenger RNA** (mRNA) (see **Table 2**). Since no **transfer RNA** (tRNA) molecules exist that possess **anticodons** to these sequences, they cannot be translated. When they occur in frame on an mRNA molecule, protein synthesis stops and the ribosome releases the mRNA and the protein. |

# T

| Telomere | End of a **chromosome**. The telomere is a specialized structure involved in replicating and stabilizing linear **DNA** molecules. |
| --- | --- |
| Teratogen | Any external agent/factor that increases the probability of congenital malformations. A teratogen may be a drug, whether prescribed or illicit, or an environmental effect, such as high temperature. The classical example is thalidomide, a drug originally prescribed for morning sickness, which resulted in very high rates of congenital malformation in exposed fetuses (especially limb defects). |
| Termination codon | See **stop codon** |
| Thymine (T) | One of the bases making up **DNA** and **RNA** (pairs with **adenine**). |
| Transcription | Synthesis of single-stranded **RNA** from a double-stranded **DNA** template (see **Figure 18**). |
| Transfer RNA (tRNA) | An **RNA** molecule that possesses an **anticodon** sequence (complementary to the **codon** in mRNA) and the amino acid which |

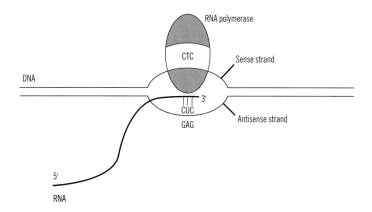

**Figure 18.** Schematic demonstrating the process of **transcription**. The sense strand has the sequence CTC (coding for leucine). **RNA** is generated by pairing with the antisense strand, which has the sequence GAG (the complement of CTC). The RNA produced is the complement of GAG, CUC (essentially the same as CTC, **uracil** replaces **thymine** in RNA).

that codon specifies. When the ribosome 'reads' the mRNA codon, the tRNA with the corresponding **anticodon** and amino acid is recruited for protein synthesis. The tRNA 'gives up' its amino acid to the production of the protein.

**Translation**

Protein synthesis directed by a specific **messenger RNA** (mRNA), (see **Figure 19**). The information in mature mRNA is converted at the ribosome into the linear arrangement of amino acids that constitutes a protein. The mRNA consists of a series of trinucleotide sequences, known as **codons**. The **start codon** is AUG, which specifies that methionine should be inserted. For each codon, except for the **stop codons** that specify the end of translation, a **transfer RNA** (tRNA) molecule exists that possesses an **anticodon** sequence (complementary to the codon in mRNA) and the amino acid which that codon specifies. The process of translation results in the sequential addition of amino acids to the growing polypeptide chain. When translation is complete, the protein is released from the ribosome/mRNA complex and may then undergo posttranslational modification, in addition to folding into its final, active, conformational shape.

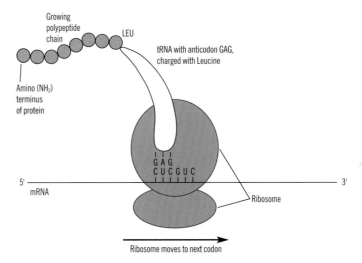

**Figure 19.** Schematic of the process of **translation**. **Messenger RNA** (mRNA) is translated at the ribosome into a growing polypeptide chain. For each **codon**, there is a **transfer RNA** (tRNA) molecule with the anticodon and the appropriate amino acid. Here, the amino acid leucine is shown being added to the polypeptide. The next codon is GUC, specifying valine. Translation happens in a 5′ to 3′ direction along the mRNA molecule. When the stop codon is reached, the polypeptide chain is released from the ribosome.

**Translocation**

Exchange of chromosomal material between two or more **nonhomologous chromosomes**. Translocations may be balanced or unbalanced. Unbalanced translocations are those that are observed in association with either a loss of genetic material, a gain, or both. As with other causes of **genomic** imbalance, there are usually phenotypic consequences, in particular mental retardation. Balanced translocations are usually associated with a normal **phenotype**, but increase the risk of genomic imbalance in offspring, with expected consequences (either severe phenotypes or lethality). Translocations are described by incorporating information about the chromosomes involved (usually but not always two) and the positions on the chromosomes at which the breaks have occurred. Thus t(11;X)(p13;q27.3) refers to an apparently balanced translocation involving chromosome 11 and X, in which the break on 11 is at 11p13 and the break on the X is at Xq27.3

| | |
|---|---|
| **Triplet repeats** | Tandem repeats in **DNA** that comprise many copies of a basic trinucleotide sequence. Of particular relevance to disorders associated with **dynamic mutations**, such as Huntington's chorea (HC). HC is associated with a pathological expansion of a CAG repeat within the coding region of the huntingtin **gene**. This repeat codes for a tract of polyglutamines in the resultant protein, and it is believed that the increase in length of the polyglutamine tract in affected individuals is toxic to cells, resulting in specific neuronal damage. |
| **Trisomy** | Possessing three copies of a particular **chromosome** instead of two. |
| **Tumor suppressor genes** | **Genes** that act to inhibit/control unrestrained growth as part of normal development. The terminology is misleading, implying that these genes function to inhibit tumor formation. The classical tumor suppressor gene is the Rb gene, which is inactivated in retinoblastoma. Unlike **oncogenes**, where a **mutation** at one **allele** is sufficient for malignant transformation in a cell (since mutations in oncogenes result in increased activity, which is unmitigated by the normal allele), both copies of a tumor suppressor gene must be inactivated in a cell for malignant transformation to proceed. Therefore, at the cellular level, tumor suppressor genes behave recessively. However, at the organismal level they behave as dominants, and an individual who possesses a mutation in only one Rb allele still has an extremely high probability of developing bilateral retinoblastomas. |

The explanation for this phenomenon was first put forward by Knudson and has come to be known as the **Knudson hypothesis** (also known as the second hit hypothesis). An individual who has a germ-line mutation in one Rb allele (and the same argument may be applied to any tumor suppressor gene) will have the mutation in every cell in his/her body. It is believed that the rate of spontaneous somatic mutation (defined functionally, in terms of loss of function of that gene by whatever mechanism) is of the order of one in a million per gene per cell division. Given that there are many more than one million retinal cells in each eye, and many cell divisions involved in retinal development, the chance that the second (wild-type) Rb allele will suffer a somatic mutation is extremely high. In a cell that has acquired a 'second hit', there will now be no functional copies of the Rb gene, as the other allele is already mutated (germ-line mutation). Such a cell will have completely lost its ability to control cell growth and will eventually manifest as a retinoblastoma. The same mechanism occurs in many other tumors, the tissue affected being related to the tissue specificity of expression of the relevant tumor suppressor gene.

# U

**Unequal crossing over**

Occurs between similar sequences on **chromosomes** that are not properly aligned. It is common where specific repeats are found and is the basis of many **microdeletion/microduplication** syndromes (see **Figure 20**).

**Uniparental disomy (UPD)**

In the vast majority of individuals, each **chromosome** of a pair is derived from a different parent. However, UPD occurs when an offspring receives both copies of a particular chromosome from only one of its parents. UPD of some chromosomes results in recognizable **phenotypes** whereas for other chromosomes there do not appear to be any phenotypic sequelae. One example of UPD is Prader–Willi syndrome (PWS), which can occur if an individual inherits both copies of chromosome 15 from their mother.

**Uniparental heterodisomy**

**Uniparental disomy** in which the two **homologues** inherited from the same parent are not identical. If the parent has **chromosomes** A,B the child will also have A,B.

**Uniparental isodisomy**

**Uniparental disomy** in which the two **homologues** inherited from the same parent are identical (i.e. duplicates). So, if the parent has **chromosomes** A,B then the child will have either A,A or B,B.

**Uracil (U)**

A nitrogenous base found in **RNA** but not in **DNA**, uracil is capable of forming a **base pair** with **adenine**.

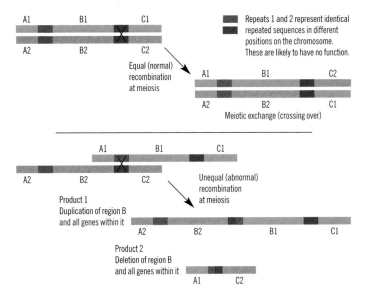

**Figure 20.** Schematic demonstrating (i) normal **homologous recombination** and (ii) homologous unequal recombination, resulting in a **deletion** and a duplication **chromosome**.

# V

**Variable expressivity**

Variable expression of a **phenotype**: not all-or-none (as is the case with **penetrance**). Individuals with identical **mutations** may manifest variable severity of symptoms, or symptoms that appear in one organ and not in another.

**Variable number of tandem repeats (VNTR)**

Certain **DNA** sequences possess tandem arrays of repeated sequences. Generally, the longer the array (i.e. the greater the number of copies of a given repeat), the more unstable the sequence, with a consequent wide variability between **alleles** (both within an individual and between individuals). Because of their variability, VNTRs are extremely useful for genetic studies as they allow for different alleles to be distinguished.

# W

**Western blot**    Like a **Southern** or **Northern blot** but for proteins, using a labeled antibody as a probe.

# X

**X-autosome translocation** **Translocation** between the **X chromosome** and an **autosome**.

**X chromosome**    See **sex chromosomes**

**X-chromosome inactivation**    See **lyonization**

**X-linked**    Relating to the **X chromosome**/associated with **genes** on the X chromosome.

**X-linked recessive (XLR)** **X-linked** disorder in which the **phenotype** is manifest in **homozygous/hemizygous** individuals (see **Figure 21**). In practice, it is hemizygous males that are affected by X-linked recessive disorders, such as Duchenne's muscular dystrophy (DMD). Females are rarely affected by XLR disorders, although a number of mechanisms have been described that predispose females to being affected, despite being **heterozygous**.

**X-linked dominant (XLD)** **X-linked** disorder that manifests in the heterozygote. XLD disorders result in manifestation of the **phenotype** in females and males (see **Figure 22**). However, because males are **hemizygous**, they are more severely affected as a rule. In some cases, the XLD disorder results in male lethality.

# Y

**Y chromosome**    See **sex chromosomes**

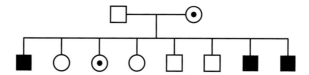

**Figure 21a. X-linked recessive inheritance** – A. Most X-linked disorders manifest recessively, in that **heterozygous** females (**carriers**) are unaffected and males, who are **hemizygous** (possess only one **X** chromosome) are affected. In this example, a carrier mother has transmitted the disorder to three of her sons. One of her daughters is also a carrier. On average, 50% of the male offspring of a carrier mother will be affected (having inherited the mutated X chromosome), and 50% will be unaffected. Similarly, 50% of daughters will be carriers and 50% will not be carriers. None of the female offspring will be affected but the carriers will carry the same risks to their offspring as their mother. The classical example of this type of inheritance is Duchenne's muscular dystrophy.

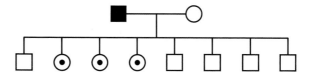

**Figure 21b. X-linked recessive inheritance** – B. In this example the father is affected. Because all his sons must have inherited their **Y chromosome** from him and their **X chromosome** from their normal mother, none will be affected. Since all his daughters must have inherited his X chromosome, all will be carriers but none affected. For this type of inheritance, it is clearly necessary that males reach reproductive age and are fertile—this is not the case with Duchenne's muscular dystrophy, which is usually fatal by the teenage years in boys. Emery-Dreifuss muscular dystrophy is a good example of this form of inheritance, as males are likely to live long enough to reproduce.

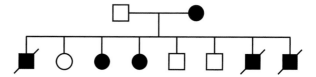

**Figure 22. X-linked dominant inheritance.** In X-linked dominant inheritance, the **heterozygous** female and hemizygous male are affected, however, the males are usually more severely affected than the females. In many cases, X-linked dominant disorders are lethal in males, resulting either in miscarriage or neonatal/infantile death. On average, 50% of all males of an affected mother will inherit the gene and be severely affected; 50% of males will be completely normal. Fifty percent of female offspring will have the same phenotype as their affected mother and the other 50% will be normal and carry no extra risk for their offspring. An example of this type of inheritance is incontinentia pigmenti, a disorder that is almost always lethal in males (males are usually lost during pregnancy).

# Z

**Zippering**　　　　　A process by which complementary **DNA** (cDNA) strands that have annealed over a short length undergo rapid full annealing along their whole length. DNA annealing is believed to occur in two main stages. A chance encounter of two strands that are complementary results in a short region of double-stranded DNA (dsDNA), which if perfectly matched, stabilizes the two single strands so that further re-annealing of their specific sequences proceeds extremely rapidly. The initial stage is known as nucleation, while the second stage is called zippering.

**Zygote**　　　　　　**Diploid** cell resulting from the union of male and female **haploid gametes**.

# Index